Understanding the Role of Business Analytics

Hardeep Chahal · Jeevan Jyoti · Jochen Wirtz
Editors

Understanding the Role of Business Analytics

Some Applications

 Springer

Editors
Hardeep Chahal
Department of Commerce
University of Jammu
Jammu, India

Jochen Wirtz
Department of Marketing
National University of Singapore
Singapore, Singapore

Jeevan Jyoti
Department of Commerce
University of Jammu
Jammu, India

ISBN 978-981-13-1333-2   ISBN 978-981-13-1334-9   (eBook)
https://doi.org/10.1007/978-981-13-1334-9

Library of Congress Control Number: 2018949885

This Springer imprint is published by the registered company Springer Nature Singapore Pte Ltd.
The registered company address is: 152 Beach Road, #21-01/04 Gateway East, Singapore 189721, Singapore

Foreword

Changes in business environment have created many opportunities as well as uncertainties, which make the role of business research more important for the decision-makers. Since managers have to make quick but accurate decisions in order to sustain the competition, business and service firms require knowledge of advanced research techniques besides the routine projections and trend analyses. Recent research studies have highlighted the growing need for the application of advanced techniques in the business decision-making process as business analytical techniques provide managers with more confidence in dealing with uncertainty despite a flood of available data. In this context, the book edited by Dr. Chahal, Dr. Jyoti, and Dr. Wirtz is an excellent initiative to present the work of eminent researchers from various parts of the world including USA, UK, France, Singapore, Iran, UAE, and India. I have gone through the manuscripts of the contributors. The authors have analysed the data with the help of various analytical techniques like exploratory and confirmatory factor analysis, regression modelling, forecasting, structural equation modelling for better information of managers to take better quality decisions. The book will equip the stakeholders including managers, practitioners, entrepreneurs, researchers, and individuals working in the extant, complex, and uncertain environment with empowered knowledge and skills to use and interpret statistical techniques for attaining sustainable, competitive advantage.

I wish success and best of luck to all who have contributed in making this initiative successful.

Greater Noida, India   Prof. R. D. Sharma
                                                       Vice-Chancellor
                                          Noida International University

Jammu, India   Former Vice-Chancellor
                                                   University of Jammu

Preface

To sustain competition in current business environment, managers have to be more decisive to offer high-quality goods/services at cost-effective rates. In addition to the routine projections and trend analyses, they require the knowledge of advanced research techniques to make quick and accurate decisions.

Data in its raw form is usually useless, and the driving force behind any data-driven organization is insights and conclusions drawn from the data, which can suggest a new course of action. In order to draw insights and reach conclusions, managers need analytical tools and techniques to interpret data from various sources and use the results for better decision-making. Further, business analytical tools also help the researchers and academicians in better theory development. Many researchers have claimed that the availability of business analytics has played a great role in converting organizations into high-performance work systems. Companies using these techniques in their decision-making process are in a better position to compete and sustain competitive advantage by minimizing risk, investing in accurate innovations, and above all providing a better picture of what is practically viable and non-viable.

The significance of the analytical needs can be judged from the fact that a significant proportion of high-performance companies have high analytical skills among their personnel. And companies employing data analytical methods and techniques in their decision-making process are in a better position to compete and sustain competitive advantage. Among the various statistical techniques, structural equation models (SEMs), including confirmatory factor analysis, help in both theory building and predictive analysis, and their roles have become more crucial with the advent of big data. Carrying out predictive modelling on large data sets has the potential to generate fresh insights for business practitioners and drive new theories for management researchers. Addressing this need, our efforts in this context are to fill the extant gap and help managers and entrepreneurs in knowledge-based decision-making.

This edited book is a collection of ten chapters covering diverse data analytics topics including a conceptual chapter on big data (Chap. 2) and eleven empirical chapters on various functional areas like finance (Chaps. 3–6), marketing (Chaps. 7 –8), and HR/OB (Chaps. 9–10). The contributors have used basic techniques like correlation, forecasting, and trend analysis and advanced higher order modelling techniques like a gravity model and a panel data quantile-regression, structural equation modelling, mediation analysis, moderation analysis etc. These chapters are going to be very useful to the researchers and practitioners in the application of analytical tools and techniques for better strategic decision-making.

Jammu, India   Hardeep Chahal
Jammu, India   Jeevan Jyoti
Singapore, Singapore   Jochen Wirtz

Acknowledgements

We acknowledge all those people who were involved and helped in completing this project. At the outset, we would like to thank the authors who have contributed to this book in terms of their time and expertise. We also wish to acknowledge the valuable contributions of the reviewers regarding the improvement of quality, coherence, and content presentation of chapters. We also appreciate the referees for reviewing the chapters, and scholars for editing and organizing the chapters.

Hardeep Chahal
Jeevan Jyoti
Jochen Wirtz

Contents

Editors and Contributors

About the Editors

Dr. Hardeep Chahal is Professor in the Department of Commerce, University of Jammu, India. Her research interests focus on services marketing with an emphasis on consumer satisfaction and loyalty, service quality, brand equity, and market orientation. Her work has been published in refereed international journals like *Managing Service Quality*, *International Journal of Health Care Quality Assurance*, *International Journal of Bank Marketing*, *Journal of Relationship Marketing*, *Journal of Health Management*, *Management Research Review*, *Total Quality Management and Business Excellence*, *Corporate Governance*, *Global Business Review*. She has also co-edited books such as "Sustainable Competitive Advantage: A Road to Success" (Excel India Publishers, New Delhi, 2015), "Research Methodology in Commerce and Management" (Anmol Publications, New Delhi, 2004), and "Strategic Service Management" (Excel Books, New Delhi, 2010). She currently serves on the editorial boards of the *International Journal of Health Care Quality Assurance* (Emerald) and *Journal of Service Research* (IIMT, India). She was Visiting Fellow at the Loughborough University, UK, under Commonwealth Fellowship Scheme (British Academy Award) and also at Gandhi Institute of Business and Technology, Jakarta, Indonesia.

Dr. Jeevan Jyoti is Assistant Professor in the Department of Commerce, University of Jammu, India, and has rich experience in teaching and research in business education. Her areas of interest are strategic human resource management, organizational behaviour, and entrepreneurship. She has publications in reputed international refereed journals such as *Personnel Review*, *Cross Cultural Management: An International Journal*, *International Journal of Management Concepts and Philosophy*, *International Journal of Educational Management*,

IIMB Management Review, *Total Quality Management and Business Excellence*, *Metamorphosis: A Journal of Management Research*, *Vision: The Journal of Business Perspective*, *Global Business Review*, *SAGE Open*. She has, to her credit, one edited book and 17 chapters in edited books.

Jochen Wirtz is Vice Dean, Graduate Studies, and Professor of marketing at the NUS Business School, National University of Singapore. He has published over 200 academic articles, chapters, and industry reports, including five features in *Harvard Business Review*. He has more than 10 books including "Services Marketing: People, Technology, Strategy" (World Scientific, eighth edition, 2016), "Essentials of Services Marketing" (Pearson Education, third edition, 2018), and "Winning in Service Markets" (World Scientific, 2017).

Contributors

Bhavna Arora Department of Computer Science & IT, Central University of Jammu, Jammu, J&K, India

Arvind Bhardwaj Department of Industrial & Production Engineering, National Institute of Technology, Jalandhar, Punjab, India

Ali Kemal Çelik Ardahan University, Ardahan, Turkey

Hardeep Chahal Department of Commerce, University of Jammu, Jammu, Jammu and Kashmir, India

Kamani Dutta University of Jammu, Jammu, India

Miraç Eren Ondokuz Mayıs University, Samsun, Turkey

Mahesh Gupta University of Louisville, Louisville, USA

S. M. Imamul Haque Department of Commerce, Aligarh Muslim University, Aligarh, UP, India

Ibrahim Huseyni Şırnak University, Şırnak, Turkey

Swatantra Kumar Jaiswal Department of Industrial & Production Engineering, National Institute of Technology, Jalandhar, Punjab, India

Jeevan Jyoti Department of Commerce, University of Jammu, Jammu, Jammu and Kashmir, India

Jagmeet Kaur Government General Zorawar Singh Memorial Degree College, Reasi, India

Phillip Klaus School of Management Centre for Advanced Research in Marketing, Cranfield University, Bedfordshire, UK

Sumeet Kour Department of Commerce, University of Jammu, Jammu, Jammu and Kashmir, India

Rahul S. Mor Department of Industrial & Production Engineering, National Institute of Technology, Jalandhar, Punjab, India

Vijay Pereira University of Wollongong, Dubai Campus, Dubai, United Arab Emirates

Shahid Hamid Raina Department of Economics, Central University of Jammu, Jammu, India

Swati Raina Lovely Professional University, Phagwara, Punjab, India

Geeta Rana Swami Rama Himalayan University, Dehradun, India

Rahul Rangotra Department of Management Studies, Central University of Kashmir, Srinagar, Jammu and Kashmir, India

Ravindra Sharma Uttarakhand Technical University, Dehradun, India; Swami Rama Himalayan University, Dehradun, India

Sarbjit Singh Department of Industrial & Production Engineering, National Institute of Technology, Jalandhar, Punjab, India

S. P. Singh Gurukul Kangri Vishwavidyalaya, Haridwar, India

Arif Ahmad Wani Department of Commerce, Aligarh Muslim University, Aligarh, UP, India

Jochen Wirtz National University of Singapore, Singapore, Singapore

Chapter 1
Business Analytics: Concept and Applications

Hardeep Chahal, Jeevan Jyoti and Jochen Wirtz

Abstract The word business analytics has become a buzzword in the present era of experience economy. Primarily, the proliferation of the Internet and information technology has made business analytics a robust application area. On the other hand, it is equally impossible to deny its significant impact on the fields of information technology, quantitative methods and the decision sciences (Cegielski and Jones-Farmer 2016). Both industry and academia seek to hire talent in these areas with the hope of developing organizational competencies to sustain competitive advantage. Hopkins et al. (2007) and Hair et al. (2003) assert that adequate knowledge on business analytics techniques enables the analysts—practitioners, managers, etc—with capabilities that enable them to take quick and smart decisions and provide stable leadership to the organization to compete in the market effectively. On the other hand, it provides a platform for the researchers and academicians to lay down path for the theory development. However, Hawley (2016) pointed that business analytics focuses more on understanding the organizational culture than mere technology. Thus, for successful implementation and harnessing the benefits of business analytics the knowledge of an organization's motivation, strengths and weaknesses is necessary (Hawley 2016).

Keywords Business analytics · Decision making · Statistical techniques Quantitative analysis · Business applications

H. Chahal · J. Jyoti (✉)
Department of Commerce, University of Jammu, Jammu, J&K, India
e-mail: jjyotigupta64@gmail.com

H. Chahal
e-mail: drhardeepchahal@gmail.com

J. Wirtz
National University of Singapore, Singapore, Singapore

1.1 Introduction

The word business analytics has become a buzzword in the present era of experience economy. Primarily, the proliferation of the Internet and information technology has made business analytics a robust application area. On the other hand, it is equally impossible to deny its significant impact on the fields of information technology, quantitative methods and the decision sciences (Cegielski and Jones-Farmer 2016). Both industry and academia seek to hire talent in these areas with the hope of developing organizational competencies to sustain competitive advantage. Hopkins et al. (2007) and Hair et al. (2003) assert that adequate knowledge on business analytics techniques enables the analysts—practitioners, managers etc—with capabilities that enable them to take quick and smart decisions and provide stable leadership to the organization to compete in the market effectively. On the other hand, it provides a platform for the researchers and academicians to lay down path for the theory development. However, Hawley (2016) pointed that business analytics focuses more on understanding the organizational culture than mere technology. Thus, for successful implementation and harnessing the benefits of business analytics the knowledge of an organization's motivation, strengths and weaknesses is necessary (Hawley 2016).

Business analytics comprises techniques and processes, namely statistical analysis; forecasting; predictive analysis and optimization, which maintain and sustain business performance (Davenport and Harris 2006; Hopkins et al. 2007). It is

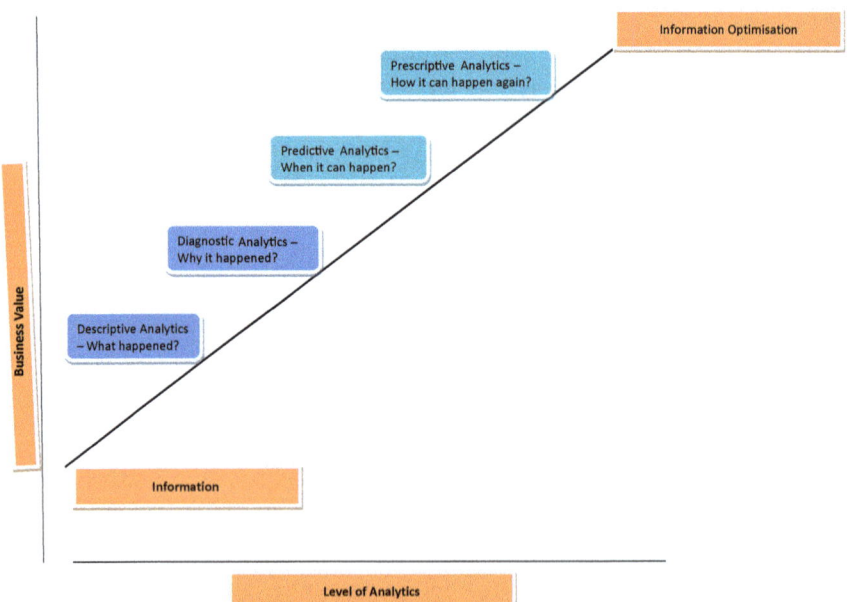

Fig. 1.1 Business analytics stages

spread across four stages—descriptive analytics, diagnostic analytics, predictive analytics and prescriptive analytics (Fig. 1.1). Each stage helps the practitioners in harnessing business value depending upon the nature of importance and business objectives. Accordingly, business analytics has a wide range of application in varied business areas that include customer relationship management, human resource management, financial management, marketing, supply chain management across all sectors. Its application facilitates and equips business and service organizations with better understanding of primary and secondary data for decision-making. The effective application of business analytics can help the business and service organizations to improve profitability, increase market share and revenue and provide better return to a shareholder.

1.2 Business Analytics: Its Applications

Research, in the last few decades, highlights on the growing need of application of advanced techniques in the business decision-making for managers (Evermann and Tate 2016). The role of business analytics has become paramount due to complex business problems, limited ability to analyse the available solutions and shortage of time for decision-making (Davenport and Harris 2006). They claim that business analytical techniques provide managers with more confidence in dealing with uncertainty in spite of the availability of huge data. Besides, these techniques are of great interest and utility to behavioural and social scientists also, who continually struggle with complex phenomenon, to detect pattern buried in complex quantitative data. Hopkins et al. (2007) and Hair et al. (2003) assert that adequate knowledge on business analytics techniques enables the analysts—practitioners, managers, etc—with capabilities that enable them to take quick and smart decisions and provide stable leadership to the organization to compete in the market effectively. On the other hand, it provides a platform for the researchers and academicians to lay down path for the theory development.

While highlighting on the significance of the analytical needs, Davenport and Harris (2006) claimed that most of the high-performance work systems/organizations have employees with high analytical capabilities. And companies using such techniques in their decision-making process are in a better position to compete and sustain competitive advantage. Among the various statistical techniques, structural equation models (SEMs), including confirmatory factor analysis, help in both theory building and predictive analysis and their role has become more crucial with the advent of big data. The predictive modelling enables managers and the researchers to have fresh insights for their future endeavours (Shmueli and Koppius 2011). Addressing this need, our efforts in this context will fill the extant gap and will help managers and entrepreneurs in a knowledge-based decision-making (Hair Jr. et al. 2011).

This edited book is a collection of ten research papers including a conceptual paper on big data (Chap. 2) and nine empirical papers in the areas of finance

(Chaps. 3–6), marketing (Chaps. 7–9) and HR/OB (Chap. 10) that can help the researchers and practitioners in the application of analytical tools and techniques for strategic decision-making.

The second chapter on big data analytics delves upon the underlying technologies used by organizations for value generation. The author has discussed the challenges faced by business organizations in monitoring the data that has grown from terabytes to exabytes and petabytes. Further, the compounded rate of data is further growing much fast. The deluge of data generated, which is both valuable and challenging, along with emerging technologies and techniques that are used to handle it is referred to as the evolution and era of "big data". She further expressed that to leverage the large volume of data for driving the business enterprises, timely and accurate insights derived out of the big data are a big challenge. Further, handling and analysis of big data are a challenge for all types of organizations with respect to its storage and technical expertise. The chapter highlights big data characteristics like volume, value, variety, velocity, veracity and variability and its analysis through exploratory, confirmatory and qualitative data analyses. Further, technologies like Hadoop and Apache Spark in handling big data have also been discussed.

The following section of the chapter discusses in brief the contribution of the papers in the three functional areas: finance; marketing and HR/OB.

1.3 Finance

Nowadays, macroeconomic models are being used to forecast the future of the economy. Modern economics and business management are using econometric applications for extensive training of its personnel. Managers are using econometric applications for devising optimal economic strategies for better insight, superior value, optimized solutions and sustain competition. Econometric applications provide organizations with a potent set of tools to unlock the power of information for effective decision-making (Kolluru and Mishra 2012). In this context, Huseyni, Celik and Eren (Chap. 3) used a gravity model approach to analyse the primary factors that influence Turkey's vehicle (car, minibus, bus, van and truck) exports to its major trading partners over the period of 2005–2015. For this purpose, a gravity model and a panel data quantile regression approaches have been performed by the bootstrap method for empirical results. The results revealed that the population of importer country and the amount of per capita income are positively correlated with the amount of Turkish automotive exports. Additionally, when the distance between importer country and the capital of Turkey increases, the amount of Turkish automotive industry exports was more likely to have a decreasing behaviour. Further, exporting to an EU member country has a statistically significant increasing impact on the amount of automotive industry exports. The estimation results also indicated that the real exchange rate was not a statistically significant determinant of the amount of automotive industry exports during the sample period. The authors concluded that Turkey cannot exactly succeed to use the competitive advantage of the possible declines on real exchange rates due to higher costs of imports in the automotive industry.

The authors of fourth chapter, Wani, Haque and Raina, empirically proved the positive correlation of gross domestic product (GDP), inflation, lending interest rate (LIR) and capital-to-risk weighted assets ratio (CRAR) with the net interest margin (NIM) in the banking industry. The study established that a favourable macroeconomic environment proves to be a main driver for encouraging NIM with a prudent control over CRAR along with NPLs. The study suggested installing the latest advances and practices of risk management especially on the credit front, which will also help the banks to utilize excessive capital rather than accumulating it unnecessarily.

In fifth chapter, Rangotra analysed the impact of various reforms undertaken by the Government of India to improve liquidity, transparency and security in the Indian bond market. It considers reforms initiated by the Government of India since 1992 that include introduction of system of primary dealers, establishment of Clearing Corporation of Indian Limited as a clearing house, introduction of screen-based trading in government securities through Negotiated Dealing System-Order Matching (NDS-OM), trading of bonds through stock exchanges, introduction of delivery versus payment system. The impact of reforms on the Indian bond market has been examined by analysing the combined gross borrowing of centre and state government through government securities (increased by around 8900% from 1991–92 to 2016–17), secondary market transactions in government securities (increased by around 430,000% from September 1994 to September 2017), net corporate debt outstanding (increased by around 225% from June 2010 to September 2017), total trade in corporate bond market (increased by around 1450% from 2007–08 to 2016–17) and other variables related to the liquidity and size of Indian bond market. The impact of reforms is found to be positive for all the dimensions but has a significant impact only on the size and liquidity of the Indian bond market. Mor, Jaiswal, Singh and Bhardwaj (Chap. 6) have focused on demand forecasting of the short lifecycle dairy products. They have compared the performances between different forecasting models for the prediction of group of dairy products. Authors compared the moving average, regression, multiple regression and Holt-Winters models based on MAPE, MAD, MSE and RMSE for the demand forecasting of a time series formed by a group of dairy products.

1.4 Marketing

Contemporary business organizations use business intelligence and analytics to solve the magnitude and impact of data-related problems (Chen et al. 2012). It is creating an exemplary change in the way data are being used, and the marketing and sales department is no exception to this. It is pivotal that marketing professionals should become tech savvy and use technology to harness the importance of business analytics (Proschoolonline 2017). In the context of marketing, business analytics integrates market and customer-related data, technology, quantitative analysis and computer-based models to provide managers with various relevant prospective for better and optimal decision-making. Among varied areas in marketing, relationship management with customers and employees is a pivotal area in the service economy that demands continuous monitoring by service firms to sustain competitive advantage.

In this context, three papers are based on primary data that encompass customers' and employees' perspectives and are analysed using advanced structural equation modelling technique to understand how it helps in decision-making for enhancing organizational performance. Raina, Klaus, Dutta and Chahal have studied the impact of customer experience on marketing outcomes in financial services. The study establishes customer experience as multidimensional construct comprising brand experience, service experience and post-purchase experience. The study has used systematic data analytical process that includes exploratory factor analysis, item analysis and confirmatory factor for construct validation. The authors have also established the relationship between customer experience and marketing outcomes (customer satisfaction, behavioural loyalty intentions, word of mouth and service value) using structural equation modelling. The result revealed that moment of truth is the most important factor that has to be considered by financial services for creating favourable customer experience quality followed by outcome focus, peace of mind and product experience. Further, product experience has low association with customer satisfaction, behavioural loyalty intentions, word of mouth and service value. The relatively weak association with all marketing outcomes suggests that customer awareness about competitive services has increased and they no more accept every type of services from the same service provider because of varied customers' choices and their ability to compare offerings with different service providers.

The study by Kaur, Chahal and Gupta (Chap. 8) has used advanced structural equation modelling and moderation techniques to re-investigate the role of market orientation and environmental turbulence in marketing capabilities and business performance. The paper has explored and established marketing capabilities as multidimensional scale using three-stage process—EFA, item analysis and CFA—that consist of: outside in, inside out and spanning dimensions and market orientation as a function of four factors relating to intelligence generation I (customers needs), intelligence generation II (customers satisfaction), intelligence dissemination and responsiveness, both of which play significant role in enhancing organizational performance. The authors used advanced marketing analytics to establish positive relationship of marketing capabilities with market orientation and organizational performance. Further using SEM-based mediation modelling approach, authors also found that marketing capabilities act as a mediating variable between market orientation and marketing capabilities and market orientation and financial performance relationships. Further using SEM-based multigroup analysis, Gupta et al. have established the moderating role of environment turbulence in marketing capabilities and market orientation relationship.

1.5   Human Resources/OB

Understanding organization and its people have gained immense attention in the present business scenario due to the value attached to human aspects for providing sustainable competitive advantage. Human resources are rare, valuable and cannot be copied or substituted (Barney 1991). These have immense creative capabilities to upgrade the innovative domain of an organization. Though cultural diversity results in knowledge sharing at various platforms in the organization, it also has adjustment issues. In this context, Kour, Jyoti and Pereira (Chap. 9) have evaluated the role of cultural adjustment (CCA) and work experience between cultural intelligence (CQ) and knowledge sharing relationship in the banking sector. Structural model explains the indirect effect of CQ on knowledge sharing with cross-cultural adjustment as mediator. Further, the role played by language proficiency and work experience has also been evaluated. The result revealed that CCA mediates the combined effect of CQ and work experience on knowledge sharing. The study contributes towards cultural intelligence theory. It helps in understanding the complex relationships in organizational setup, which can be of immense use for the practitioners at the workplace. The last chapter by Kumar, Singh and Rana analyses the impact of employer branding on organizational attractiveness in Indian organizations using factor analysis, Pearson's r and step-wise multiple regression analysis techniques. The results indicate that employer branding has a positive and significant relationship with organizational attractiveness. Since economic value, application value, social value and development value emerged as strong predictors of attracting and retaining employees, employers can provide employees with marketable skills through training and development in return for effort and flexibility. The authors believed that the study findings can help in identifying varied EBs aspects that are effective in extracting organizational attractiveness and incorporating them into the organizational culture.

There is significant evidence that the ability to make better decisions improves with the usage of quantitative-, qualitative- and financial-based techniques. Hence, this book offers a relevant resource that can help the audience (research scholars, practitioners, market researchers, etc.) in the application and interpretation of statistical practices, using real-world applications from the fields of marketing, human resources, finance, operations research and information technology relating to issues like preferences of a customer base, quality of manufactured products, high-performance human resource policy, employee resilience, availability of financial resources, operational flexibility, etc.

References

Barney, J. (1991). Firm resources and competitive advantage. *Journal of Management, 17*(1), 99–120.

Cegielski, C. G., & Jones-Farmer, L. A. (2016). Knowledge, skills, and abilities for entry-level business analytics positions: A multi-method study. *Decision Sciences Journal of Innovative Education, 4*(1), 91–118.

Chen, H., Chiang, R. H., & Storey, V. C. (2012). Business intelligence and analytics: From big data to big impact. *MIS Quarterly, 36*(4), 1165–1188.

Davenport, T. H., & Harris, J. G. (2006). *Competing on analytics: The new science of winning.* Boston, MA: Harward Business School Press.

Evermann, J., & Tate, M. (2016). Assessing the predictive performance of structural equation model estimators. *Journal of Business Research, 69,* 4565–4582.

Hair, J. F., Bush, R. P., & Ortinau, D. J. (2003). *Marketing research: Within a changing information environment* (2nd ed.). New York: McGraw-Hill/ Irwin.

Hair, J. F., Jr., Wolfinbarger, M., Money, A. H., Samouel, P., & Page, M. J. (2011). *Essentials of Business Research Methods.* Armonk, N.Y.: M.E. Sharpe.

Hawley, D. (2016). Implementing business analytics within the supply chain: success. *The Electronic Journal Information Systems Evaluation, 19*(2), 112–120.

Hopkins, M. S., LaValle, S., Balboni, F., Kruschwitz, N., & Shockley, R. (2007). 10 insights: A first look at the new intelligence enterprise survey on winning with data. *MIT Sloan Management Review, 52*(1), 21–31.

Kolluru, S., & Mishra, R. K., (2012). *Econometric Applications for Managers.* New Delhi: Allied Publisher.

Proschoolonline. (2017, June). Proschoolonline.com/blog. Retrieved 2017, from Business-analytics-for-marketing-professionals: http://www.proschoolonline.com/blog/business-analytics-for-marketing-professionals/.

Shmueli, G., & Koppius, O. R. (2011). Predictive analytics in information system. *MIS Quarterly, 35*(3), 553–572.

Chapter 2
Big Data Analytics: The Underlying Technologies Used by Organizations for Value Generation

Bhavna Arora

Abstract The expansion of Internet and its applications globally has witnessed generation of high volume of data resulting in high volume of information. In the contemporary era of digital world, data is seen as the driving force behind the progression of business enterprises. Today, the data that is generated worldwide has grown ranging from terabytes to exabytes and petabytes, and the compounded rate of data further growing is much fast. The data generated widely has many forms and structures. The deluge of data generated, which is both valuable and challenging, along with emerging technologies and techniques that are used to handle it is referred to as the evolution and era of "Big Data". As the big data is generated from multitudinous sources, majority of this data exists in unstructured form that demands specialized processing and storage capabilities, unlike the structured data that uses storage and processing of traditional relational structures. This results in high complexity and uncertainty in data. The usage of statistical analysis, computer-based models and quantitative methods that can help the business organizations to improve insights for better operations and decision-making is referred as business analytics. To work intelligently and focus on value generation, organizations need to focus on business analytics. The analytics are a critical component of big data computing. As defined in the literature, an intelligent enterprise has the characteristics similar to human nervous system and is responsive to external stimuli. To leverage the large volume of data for driving the business enterprises, timely and accurate insights derived out of the big data are a big challenge. The technologies like Hadoop and Apache Spark assist in handling big data on both fronts. However, handling and analysis of big data are a challenge for any organization with respect to its storage and technical expertise. Business analytics is used in business organizations for value generation by data manipulation along with business intelligence and report generation. Advanced analytics are also used by business enterprises that use techniques of data mining, data optimization and predictive forecasting.

B. Arora (✉)
Department of Computer Science & IT, Central University of Jammu, Jammu, J&K, India
e-mail: bhavna.aroramakin@gmail.com

© Springer Nature Singapore Pte Ltd. 2019
H. Chahal et al. (eds.), *Understanding the Role of Business Analytics*,
https://doi.org/10.1007/978-981-13-1334-9_2

Keywords Big Data · Data Analytics · Hadoop · V's of Big Data
Apache Spark

2.1 Introduction to Big Data

The contemporary era has witnessed very large volumes of data and the termi-
nology and trends that have been accepted globally with these are "Big Data". The
author in paper ("What Is Big Data?—Gartner IT Glossary—Big Data", n.d.) has
defined big data as "Big data is high-volume, high-velocity and high-variety
information assets that demand cost-effective, innovative forms of information
processing for enhanced insight and decision-making".

In Manyika et al. (2011), the author refers "Big Data" to "data set whose size is
beyond the ability of typical database software tools to capture, store, manage and
analyse".

The volume of data that is being stored today around the world is exploding. In
the year 2000, the world witnessed storage of 8 lac petabytes of data. With the
expansion of Web and its applications, the data that is being stored is growing
exponentially. The data is likely to rise to 35 zettabytes by the year 2020. The data
that is created is not analysed efficiently, and the insights of the data is not revealed.
The data contains hidden insights that the companies can use to enhance their
business perspectives. How the volume of big data impacts the human mind is very
challenging. Considering the data volume that consists of multiples of terabytes
may be considered as big data, but actually when it can be managed in network
attached storages (NAS) or storage area network (SAN) using additional disc
arrays, then it might not be considered as really Big Data. When the data exceeds
this limit, i.e. about petabytes in size and can only be managed with sophisticated
applications and tools, the data can be referred as "Big Data". This would require a
complex distributed computing and storage grids extensively, so that this data could
be managed.

However, companies used various tools and technologies to collect and store
different types of big data. The analysis of these diversities of big data is challenging as the tools that are required for big data analysis are extremely complex to
design and implement. The management of this data is another big challenge as the
companies should have the clarity on big data adoptions. It is paramount in agreeing
that such information in big data which is huge and complex has created various
challenges for organizations that did not exist earlier. The large volumes of data
available pose several problems for researchers, analysts and decision-makers in the
industry. At times, the decision-makers in the organization tend to make their
decisions without having complete facts, and others find the business intelligence
along with the data analytics to be part of their visionary plans so as to enhance
business competitiveness.

The evolution of big data has witnessed the explosive growth in the entire
world's data that can be used to make decisions, but this can only be useful if this

can be made in timely manner. For this, powerful tools are needed that can assist in storage, extraction and analysing the data from the big data sets. The big data can also be defined as that data that cannot be processed through conventional methods of processing. To mention few varieties and sources of data that come under the big data realm are as follows (Jain 2015):

1. Black box data captures the voice and recordings of flight crew members of helicopter and other aircraft along with the information pertaining to the performance of the aircraft.
2. Data of stock holdings where the decisions made by a customer on a share or equity of different companies.
3. Data from social media websites such as Facebook and Twitter that holds views and other information posted by millions of users across the globe.
4. Transport and meteorological data sources.
5. Data retrieved by search engines from different databases.
6. Metamorphic and Census data.
7. Connection-oriented data that includes sensory data.
8. Data from cloud storages that provides computing and data on demand.

Big data is more than just more information; it represents the beginning of the end of the industry experience as a core competitive advantage (Stubbs 2014). Big data is not a philosophical fancy anymore. It is already in place in industry. Big data cannot be argued as just the latest version of "data". Today, the users are generating much more data and more types of data than before. In their work, Manyika et al. (2011) have proposed five major contributions that big data contributes to business organizations:

- transparency creation by making big data more accessible and ready to use in timely manner for value generation
- performance improvement by enabling experimentation
- population segmentation by tailoring products and services that meet specific needs
- decision-making support
- innovative business models, products and services

Big data and business analytics work hand in glove. Without data, analysis cannot be done. Without business analytics, big data is just noise. Big data bears the potential of making things efficient and is capable of generating returns. These returns include benefits to internal value such as productivity or external value like revenue generation. It offers exceptional insights along with predictive capabilities for those who are able to leverage it.

2.2   The V's of the Big Data

The contemporaneous era is witnessing production of data at astronomical rates. To analyse this data that is constantly varying, new tools and technologies are continuously being developed by experts that will be able to handle the complexities of large volumes of data. The future trend is that the big data is going to grow more rather than decrease, as more and more data generating applications are growing. Big data can be characterized by volume, value, variety, velocity (Philip Chen and Zhang 2014), veracity and variability, and each of these parameters can be defined as under:

2.2.1   *Volume*

Today, the large volume of data is generated because nowadays organizations collect and process data from a diverse range of sources such as application generated logs, machine-generated data, email data, weather and geographic information systems (GIS) data, survey data, reports, social media data. Big data analytics have the capability to compute gigantic volume of information. Data volumes have reached levels to terabytes (TB) or petabytes (PB). As an example, the financial industry produces voluminous data in terms of market data, quotes and financial trading. The New York Stock Exchange creates about more than one terabyte of data per day ("Want to make big bucks in stock market? Use Big Data Analytics", n.d.) and if this volume is calculated over the month, year and so on, the volume of data is immense. About 10 billion photographs (Beaver 2008) were hosted by Facebook creating about one petabyte of data storage in the year 2008. Another site Ancestry.com, stores around four petabytes of data (Bertolucci 2013). Even the Internet archive stores about two petabytes of data, and it is accelerated at a rate of about twenty terabytes per month ("1. Meet Hadoop—Hadoop: The Definitive Guide, 3rd Edition [Book]", n.d.). These big data structures comprising of high volumes of data tends to impose limitation on the storage and processing capabilities. It also imposes limitations to the database structures, and hence, the database modelling gets complicated as the data grows. In his work (Brock and Khan 2017), the author analyses that the huge amount of data poses challenges to underlying storage infrastructure which in turns calls for systems with scalability and capability for distributed querying. High computational power and parallel processing are required for analysing big data as the traditional database techniques are not able to cope with big data, as the size of data sets has surpassed the capabilities of computation and storage.

2.2.2 Variety

The traditional systems heavily rely on underlying structured data whose dimensions are considered as accuracy, completeness, relevance and timeliness. The inputs to such systems need to be entered judiciously and meticulously so that the output that is produced in the form of reports is meaningful and useful. The big data is heterogeneous. In such environments, the system needs to use techniques for data cleaning so as to eliminate the garbage data in source. Data collected from various sources like applications, stock data, emails, geographical data, weather data, social media application data is difficult to handle as it comes from a variety of sources, and it is virtually impossible to convert this heterogeneous data to a conventional structured form for processing. In order to process such data, special techniques and technologies are used that can understand and go beyond the traditional processing of the relational structured data. Big data solutions need different types of processing tools to process heterogeneous data.

2.2.3 Velocity

The rate at which the data flows in the system and its environment is termed as velocity. With the Internet and mobile data coming in the lives of the consumer, the era witnesses high rate of data flow as the consumers carry with their devices, a huge volume of streaming source of data that consist of geo-located images and audios. Studies reveal that in the year 2013, about five exabyte data were generated in the world every 10 min. Today, this figure has risen exponentially risen, and the data is being generated every minute. However, the importance of velocity of big data follows the similar rate of increase as in the case of volume of the data. For example, the business Walmart creates about 2.5 petabytes per hour (Brock and Khan 2017). One of the main challenges to the velocity is the communication networks. Since the big data processing demands real-time processing, the processing capabilities for inflow of the data streams in the networks are also a big challenge.

2.2.4 Veracity

The reliability and trustworthiness of data are termed as veracity of data. It also refers to the quality of the data. The point to focus on is "how accurate is all this data?" As an example, consider the tweets in Twitter posts. These posts contain hashtags, typos, abbreviations, etc. The data that is to be considered should be reliable, accurate and trustworthy. Manipulation and analysis of such data need to be qualitative and trustworthy to get correct insights from it. For real-time applications, to provide correct and reliable data at times the applications may produce nearest best results in the cases where the data that is being analysed in real-time applications fails to deliver in a particular moment.

2.2.5 *Value*

Value refers to the worthiness of the data being extracted. On one hand, if the organization has voluminous amount of data but unless it can be made useful for the organization, it is worthless. Even though there is an explicit association between data and insights, it does not definitely mean that there is value in big data. The original data received might have low value as compared to its volume. By analysing a large volume of data appropriately, high value and pronounced insights from the data can be obtained. The significance of embarking on initiatives of big data is to understand the costs of analysing and reaping the benefits during the process of collecting and analysing data. Thus, it ensures that the data that is reaped is monetized and the organization is benefitted to the maximum.

2.2.6 *Variability*

The variation in the flow rates of data is referred to as variability of data. The velocity of big data is inconsistent and has periodic troughs and peaks. As the big data is generated from myriad resources, complexity also has to be analysed. Complexity arises after collecting data from different sources as it has to connect, clean, match and transform the data (Fig. 2.1).

2.3 Big Data Classification

As the data sources of big data are numerous, based on the types of data, big data can be classified broadly into three categories—unstructured, semi-structured and structured.

2.3.1 *Structured Data*

The data that is stored in the databases in an orderly manner is referred to as the structured data. Statistics reveal that structured data only constitutes about

Fig. 2.1 6V's of Big Data

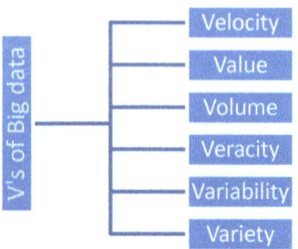

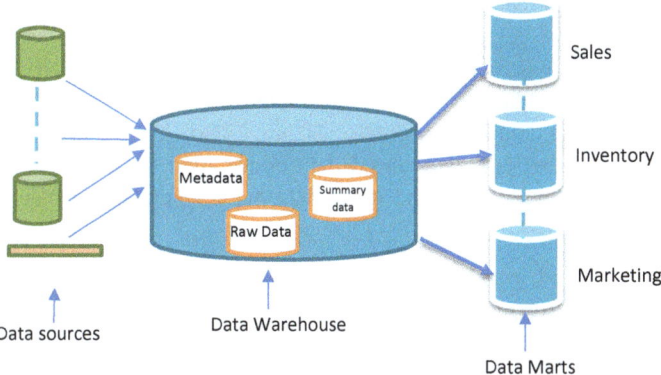

Fig. 2.2 Overview of data warehouse and data marts

one-fourth of the data that the organizations use. This data can be used in pro-gramming and querying structures from databases. The sources of such data can be from machine input or humans. Data that is generated by machines includes data from GPS-like devices or sensory devices like the medical equipment, Web interfaces and logs. Data that is included by human systems includes database records which involve human intervention in generating data. The two most popular approaches that can be used to manage large data sets in a structured way are data warehouses and its subsets, i.e. data marts. A data warehouse can be seen as an assemblage of data which is isolated from the operational systems and the decision-making process in any organization. It is a huge repository of historic data. The data is compiled and assembled from various resources so as to provide timely and accurate information. The data in a data warehouse is the extracted information from various functional units of an organization. Before the data is integrated into the warehouse, it undergoes a series of processes. These preprocesses are data cleaning, data transformation and data cataloguing. After the preprocessing, the data is available for higher-level online data mining functions. These warehouses are controlled by a centralized unit. The subset of data warehouse is a data mart. The data marts focus on specific functional area only. The warehouse and data mart both primarily vary in their scope and usage area. The structured data forms only a small subset of the Big data that is ready for analysis (Fig. 2.2).

2.3.2 Unstructured Data

Unlike the traditional row–column database structure, the unstructured data has no clear formats in storage. Such data constitutes about 80% of the total data that is included in big data. Till few years back, such data has been stored and analysed manually as it was quite difficult to analyse this data. The unstructured data can comprise the machine-generated or the human-generated data. Typical examples of unstructured machine-generated data include satellite images, data captured from radars, whereas the unstructured human-generated data is spread across the entire globe which includes data from the Web content, social media and mobile data. Users tend to upload data on Facebook, Instagram, audios and videos, etc.; they all are a part of the massive unstructured data contributed by the user. The largest unstructured data component is the video data that constitutes big data.

2.3.3   Semi-structured Data

A very thin line exists between the unstructured and semi-structured data. Unless clearly defined, the semi-structured data can appear as unstructured. Even though that the information is not typically arranged in traditional database structured formats, but in order to process this data, some properties that make the data processing easier and convenient to process are contained in it.

2.4   Data at Rest and Data in Motion

Technically speaking, in order to gain insights from big data, right technology for various processes like collecting, managing and analysing is required. The predictive purpose for the same is quite critical. Since the data is from a variety of sources and is of different types, there is a requirement of different computing platforms that support in providing meaningful insights. In order to determine the technology and processes that are required to glean the insights from the big data, it is imperative to understand the difference between data at rest and data in motion.

2.4.1   Data at Rest

In data handling systems, data at rest refers to data that is stored in stable systems. The data at rest comprises data compiled and stored in structures like spreadsheets, databases, warehouses, archives and backups, mobile data. There can be different points where data is analysed and where its action is taken. This can occur at two separate times. For example, present month's business activities can be predicted based on last month's sales data by a retailer. The action is making strategic decisions, and the activity is the sales of the previous month. Market campaigns and strategies can be planned accordingly based on the variables like customer behaviour, sale schemes. Looking at this data analysis, the business can take advantage and such decisions can impact the sales in the stores, while the customer would be benefitted with the sale schemes the store offers.

2.4.2 Data in Motion

There is a difference in the analytics for data in motion, even though the process of data collection is similar to that of data at rest. Unlike data at rest, analytics can occur at the same time when the event occurs, i.e. in real time. The data in motion refers to data that is moving from one place to another. In such cases, many different networks can be used, e.g. sending email on Internet. Many nodes are connected to same network, and the transfer of the email has to go through multiple nodes in a network. Security issues arise for data in motion, and the data needs to be protected. Another example would be locating clients and their choices at various outlets of a water park. Latency also becomes a key concern as the lag in processing may affect the business results by missing an opportunity. This type of data is also referred to as data in transit or data in flight. Data at rest and data in motion can provide quiet meaningful insights for business analysis. It is important that appropriate processing methods and infrastructure may be used and deployed in order to obtain the perfect analysis of data.

2.5 Data Analytics

The process and techniques used for examining the data with the aim and purpose for inferring and to draw conclusions about that information are data analytics. It finds its usage in business organizations to help them make better decisions. Data analytics can be distinguished from data mining with respect to the purpose, scope and focus of the analysis. In order to mine data from the huge set of data that is in question, sophisticated software is used that rely on algorithms and are capable of working on large data sets. They can uncover undiscovered patterns and hence are capable of establishing the hidden relationships. This process is referred to as data mining. Appropriate analysis of big data can help a company to achieve cost reductions and dramatic growth. So the business houses should not wait too long to exploit the potential of analytics. Big data analytics focus on inferences, i.e. the deriving to conclusion based on known facts. The analysis can be categorized as under ("Data mining versus data analysis and analytics—Fraud and fraud detection —Academic library—free online college e textbooks", n.d.):

- Exploratory data analysis (EDA)—It is the preliminary stage where the data is explored and new features are discovered.
- Confirmatory data analysis (CDA)—Existing hypotheses are proven true or false.
- Qualitative data analysis (QDA)—It is analysis of the quality parameters of data to draw conclusions from non-quantitative and non-numerical data like pictures, audios, words or text, videos.

Big data analytics finds its significance in the cases of audits when the information systems of business organizations' along with other operations, procedures and processes are under reference. Data analysis also helps to determine if the systems under reference can effectively protect data while operating efficiently and also helps the organization accomplish overall goals. Business intelligence defines analytics from various perspectives. In call centre applications, it can be defined from online analytical processing (OLAP) to customer relationship management analytics (CRM). CRM analytics includes all process that analyses data about customers and presents it to facilitate and streamline for better business decisions of the organization.

2.5.1 Types of Business Analytics

The key sub-processes defined in big data are data management and data analytics. The process of data management looks after the acquiring, storing, retrieval and preparation for data analytics. The underlying technologies for data analytics are acquisition, annotation, aggregation, etc. (Saravanakumar and Nandini 2017). For analysing various types of structured, unstructured and semi-structured data, the following analytics are used (Saravanakumar and Nandini 2017):

- **Text Analytics**: It is also known as text mining from the textual data. It refers to the extraction of high-quality information from textual data by using statistical patterns. It includes machine learning and statistical analysis of text data using techniques such as information extraction (IE), summarization text, question answering and sentiment analysis. Tools for text analytics are SAS text analytics, IBM text analytics, SAP text analytics, etc.
- **Audio Analytics**: To analyse the unstructured audio data, speech or audio analytics is used. In audio analytics, information is extracted from natural language, i.e. languages spoken by humans. The most popular application for audio analytics is the call centres which have data for million hours and can be used to improve the customer experience and to enhance the business turnover. For audio analytics, two approaches are used—transcript-based approach and phonetic-based approach. The tools that are used for audio analytics are Marsyas, Vamp, SoundRuler and WaveSurfer, etc.
- **Video Analytics**: To monitor, detect and analyse data from video streams is referred to video analytics. This includes determining meaningful data from temporal and spatial events. The key applications where video analytics help are retail stores, health centres, transportation, securities, etc. Video analytics is also called video content analysis. This technology uses CCTV and surveillance cameras for detecting breaches, recognizing suspicious activities, etc. The tools used are Ooyala, Vidyard, Vimeo Analytics, etc.
- **Social Media Analytics**: The social media data consists of information that is gathered from websites such as Facebook, Twitter and blogs. The data needs to

be analysed by business houses for decision-making by studying behaviours and pattern of the user. The user opinions are extracted and analysed. The analytics can be content based and structured based for social media analytics. Tools used are ViralWoot, Collecto, SumAll, Tailwind, Beevolve, etc.

- **Predictive Analytics**: Historical and present data is analysed, and based on this, data prediction is made. It expresses reliability that what might happen in the future. This method is based on statistical methods. The various tools to perform predictive analytics are splunk, medalogux, etc.

2.6   Big Data Paradigm in Business Organizations

Since past few years, various business enterprises and other organizations are storing a large amount of data in large databases in data warehouses and data marts. However, data was analysed with data-mining algorithms to extract insights. Nowadays, the data stored is no longer homogenous in nature, but on contrary, it is a compilation from a variety of sources. The data in the traditional systems was organized and structured in rows and columns as it was largely generated from transactions. On the contrary, nowadays, the stored data is unstructured and generated from a variety of sources like audio–videos, photographs, text messages, maps generated from GPS devices, data from emails, social media sites, etc. All these data when stored in digital media is unstructured as there cannot be a common structure that can be defined for such data. Another key characteristic of such data is its real-time accessibility. Data can be retrieved about activities and events in real time and will also influence its outcomes. It is only possible if we have an organization that is designed to operate in real time. The process design should able to analyse and use real-time data. It should be able to produce instant insights and process those insights to support real-time decisions. The "real-time" factor affects the organizations to take timely and appropriate action. Summarizing, in order to gain maximum benefit out of big data, the organizations must work in real time.

2.6.1   Business Analytics: The Organizational Transformation

Business analytics can be defined as the use of the data-driven insights to generate value in real time. It is done by understanding the business relevancy, organizational insights, performance and value measurements (Stubbs 2014). The data-driven insights include data manipulation, reporting and business intelligence and advanced analytics. The advance analytics is that form of analytics that help provide answers to questions like what happened, what will happen, why it happened and what best possible one could do (Stubbs 2014). The advanced analytics include data-driven insights that include data mining, optimization and forecasting.

It can also be defined as the deep analysis of data or content by using appropriate technologies, tools and different techniques that are typically beyond those that are used with the traditional systems. The business intelligence (BI) may be used to uncover patterns that assist in discovering deeper insights, along with generating recommendations and making predictions. The techniques of advanced analytic may include text and data mining, machine learning and pattern matching, visualization, semantic and sentiment analysis, forecasting, network and cluster analysis, multivariate statistics, graph analysis, simulation, complex event processing, neural networks ("Business Intelligence—BI—Gartner IT Glossary", n.d.).

The output of business analytics is seen as value generation in an organization. This could be internal or external (Stubbs 2014). Internal value is from the perspective of teams that are within an organization. The outside or external value is seen from outside the organization. The organization needs to create these values through its key resources, i.e. people, processes, data and the technologies. A series of activities that can be linked to achieve an outcome is defined as a process. The processes can be strongly or weakly defined. A series of specific steps that is repeatable and may be automated is strongly defined process. On the contrary, an undefined process that relies on the capability of the personnel for execution of the process to complete it successfully is a weakly defined process. To generate new assets, various tools and technologies are applied and are consolidated to a common analytical platform. The key to business analytics is facilitating change, not driving towards better outcomes. There is a major paradigm shift in the way organizations execute their operational, tactical and strategic objectives as an outcome of business analytics.

2.6.2 The Intelligent Enterprise

Irrespective of how the people act and react to situations in the organizations, most of the organizations can be seen united under a common objective. A truly intelligent enterprise operates like our nervous system (Stubbs 2014) and possesses properties of agility, adaptability, flexibility and is appropriately responsive to external stimuli. There are different levels that are described as progressive for any organization. These are the approaches usually opted by organizations for building capability. The first level is the unstructured mode; the second level is the structured mode. The real process and mode start at level three, and from here, the system starts its "best practices" from the theory of "things working". When the gap between these two is closed or reduced, the business system becomes an intelligence enterprise.

Level 1: The Unstructured Mode In this system, everyone is working hard without a plan or clarity of work. Quality is hard to measure at this level as whatever is archived is not because of design and planning but just because of the efforts of motivated individuals. Technologically, the analysts use tools that are basic desktop-centric and devote considerable amount of time to try to manage, source and exchange data within the tools that are semi-compatible. The data fragmentation is done at this level, and each of the team that works in this environment creates their own data repository. They tend to restart from beginning every time they work on a new project. However, difficulty may arise as the processes are manual, undefined, and may require substantial efforts to execute.

Level 2: Structured System Structured systems are the next level of the unstructured system. It is when the system follows higher-order patterns, but the system behaves randomly around a broader pattern. The organizations at this state try to balance local choice while considering global requirements. Constraints are set, functional and divisional strategies are established, and the entire organization tries to comply with them. Unconscious ignorance is one of the key barriers to success at this level. Data at this level exists in tabular format on networks that allow sharing of the data as well. Simple and common tools for desktop processing are used. Processes are weakly defined, and the skills are used of the employees across various processes.

Level 3–5: Towards an Intelligence Enterprise The level next to the structured system is the intelligent enterprise. This is the time when the organizations recognize the business analytics as integrated soul of the organization. The level at which the intelligent enterprises reach this point is totally process-centric. The three levels at which the intelligent enterprises work are at the team, department and enterprise levels. Understanding that business analytics is a journey and must be incorporated in all its functions and processes which is the key to success for the organization.

2.6.3 Intelligent Analysis

The pyramid hierarchy that exists in organizations suffers a number of challenges. The top management, i.e. where the strategic decisions are to be made, is far off than the actual scene where the actions are done. This time lag leads to delay in the decisions. The strategic decisions have a significant impact on various other levels of organizations. The decisions range from resource allocation to affecting the impact of the organization's competitiveness in the marketplace. Then is the level of tactical management, where the mid-level managers operate and affect the key operations like marketing, accounting, production. The focus is not on the entire organization and has lower resource implications as compared to the level above. The transaction processing system that is used by the operational management handles the structured data. It provides operational-level support. Figure 2.3 shows different levels of management and their data usage in business organizations.

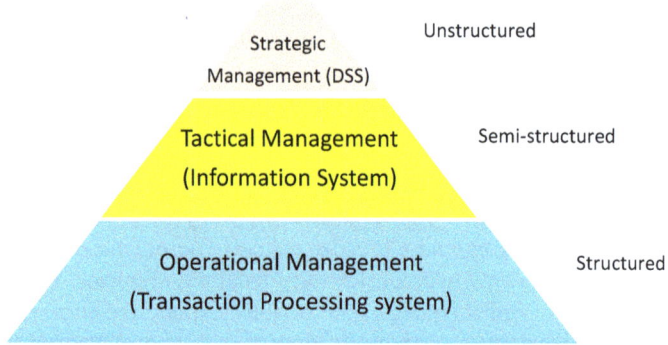

Fig. 2.3 Management system pyramid (Management systems with type of data handled)

2.7 Technologies for Data Analytics

Analytics refer to the discovery of meaningful patterns of data, e.g. data related to sales, transaction, revenue. Most information system deploy traditional database tools for relational databases such as structured query language (SQL). Business houses need data experts that have broader and deeper analytical skills that can provide support to challenges like data management, real-time analysis, real-time predictive analysis, data management issues including security and privacy (Miller 2014). The data engineer responsible for the data extraction and analysis has thorough and clear understanding of traditional relational databases along with non-traditional and NoSQL databases like Hadoop. These engineers are capable of integrating data from variety of data sources and are able to design data-driven services. They work in coordination with the scientist that works on handling of data.

In order to cope with the trends of big data, a variety of tools, techniques, methods, and technologies have been developed in recent years. When data derives in huge magnitudes, then the companies cannot rely on the in-house storage and processing anymore, as the "traditional" technology that focuses only around the central databases is no longer appropriate to handle it.

To determine what is needed and what fits in well, the requirements for big data processing need to be reviewed (Vossen 2014). These requirements can be characterized as follows:

- High processing capabilities
- High storage capabilities
- Scalability and support for distributed processing
- Fault-tolerant processing capabilities
- Support for parallel programming and processing paradigms
- Appropriate platform and execution environments.

2.7.1 Hadoop—The Underlying Technology for Big Data Analytics

The Apache Hadoop is Java-based software platform that supports data-intensive distributed applications (Philip Chen and Zhang 2014). It has been designed to avoid the low performance and the complexity encountered when processing and analysing Big data using traditional technologies (Oussous et al. 2017). The Hadoop platform is used for distributing computing and spreads the data and its processing across a number of servers. The paradigm that is used by Hadoop is the MapReduce (Fig. 2.4).

The kernel, Hadoop distributed file system (HDFS) MapReduce along with add-on projects that include Apache Hive and Apache HBase constitute the Apache Hadoop. The model of the MapReduce is used for programming and execution. It is also capable of processing and generating large volume of data sets. The underlying algorithm that is used by MapReduce is divide and conquer. The divide and conquer algorithm works by breaking a complex high-level problem into several sub-problems recursively. The sub-problems are then allocated to a cluster of working notes which solve these problems separately and in parallel. Later, the solutions are merged to give a solution to the problem in question. The Map step and the Reduce step are the two steps that are used to implement this algorithm. The two nodes that the Hadoop works on are master nodes and worker nodes. The role of the master nodes is to take the input and divide it into smaller sub-problems which are further distributed to worker nodes. Finally, the master node collects the solutions to all of the allocated sub-problems and aggregates them to produce output in Reduce step.

By default, Hadoop uses Hadoop distributed file system (HDFS). Hadoop also has the capability of working on other file systems as well. The HDFS uses the storage cluster arrays to hold the actual data. The data is dumped in HDFS, and it can be analysed within Hadoop, or it can export the data to other tools for performing analysis. The patterns used by Hadoop have three stages:

* LOAD—data into HDFS
* OPERATE—Map and Reduce sub-operations
* RETRIEVE—retrieve results from HDFS

The entire process is a batch operation. This is most suitable for analytical or non-interactive tasks. Hadoop cannot be termed as a data warehouse solution, neither it is a database, but it can support the analytical processes of the data. As an example, Facebook is the most popular application that follows the patterns of Hadoop. A database like MySQL stores the data, and this data is then replicated in Hadoop for computations and analysis (Fig. 2.5).

Fig. 2.4 Hadoop system

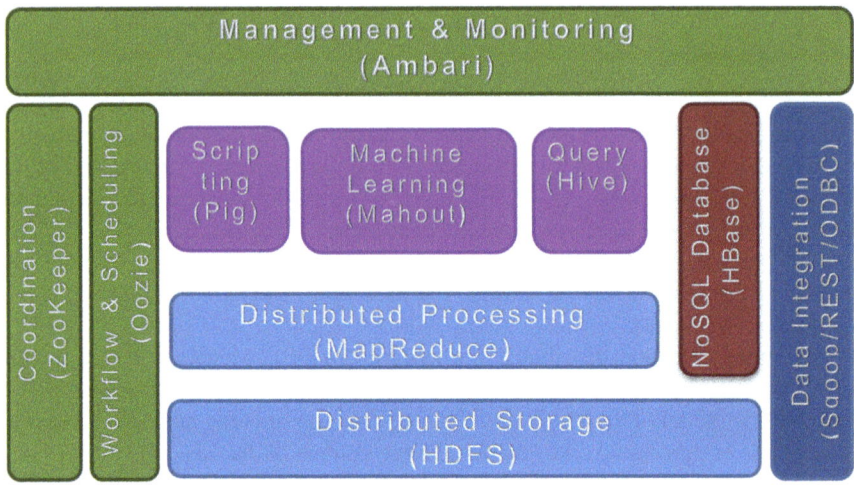

Fig. 2.5 Hadoop Ecosystem (Al-Barznji and Atanassov 2017)

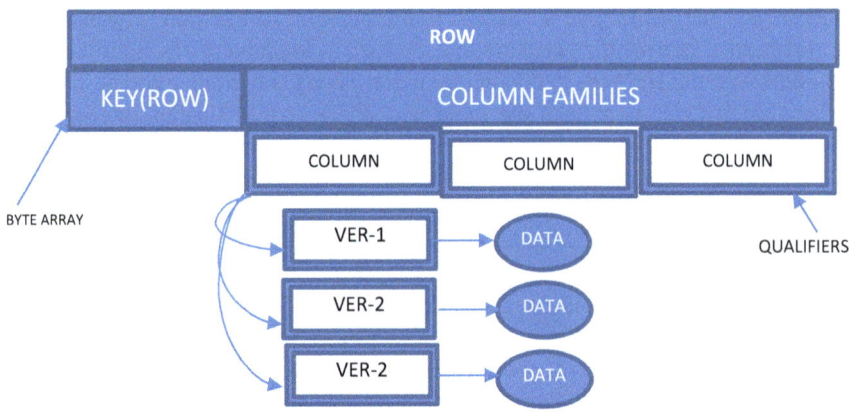

Fig. 2.6 Structure of HBase (Haines 2014)

HBase It is the Hadoop database, a NoSQL database that runs on Hadoop. It runs on the HD file system (HDFS) and provides scalability and real-time data access. This is provided as a key-value store along with the analytic capabilities of MapReduce. As the HBase is not a traditional relational database structure, it uses different methodology to model data. A four-dimensional data model is proposed for HBase. Each dimension is defined as under (Haines 2014), and the following four coordinates define each cell (Fig. 2.6):

- **Row Key**: Each of the rows in HBase has a unique key termed as row key. It is a byte array without a data type.

- **Column Family**: Data in the rows are structured into column families with every row having the same set of column families. HBase stores column families in its own data files. Any changes to be made to column families are difficult to incorporate; therefore, they need to be cautiously defined.
- **Column Qualifier**: The actual columns are referred as column qualifiers. Spread across different rows, the same column families do not require the same column qualifiers.
- **Version**: A configurable number of versions can be associated with each column. Data can be accessed for a specific version for a qualifier.

2.7.2 Hadoop in Business Organizations

The usage of Hadoop in business organizations can be understood in this example. The production server of a company stores a data set that deals with the structure and the business dealings of the company. This server facilitates the copying of the data set to an analytics engine. This can be a Hadoop cluster and can assist in the analytics of the data set. In order to prepare data and to copy it to be ready for analytics, three major processes are used. These are extract, transform and load (ETL) processes. The traditional ETL process consumes a lot of network resources along with the processing power and the bandwidth. It is noted that about 80% of the time is consumed by ETL process from each of the analytical job. Accordingly, traditional ETL may lead to excessive resource consumption and/or prolonged processing times in connection with analytics jobs.

Many organizations expand the data generated by the internal sources such as sales and services with the external demographics and social media, using the Hadoop-based analytics (Hortonworks 2013). These focus on the following key issues:

- How to identify new customer segments
- How to personalize offers
- How to reduce the customer roil

The Hadoop-based analytics can also help in businesses by reducing maintenance costs and improve asset utilization in asset-intensive industries, such as utilities, oil and gas, and industrial manufacturing. The machine-generated data along with the internally generated service data and external data can be integrated and used for predictive analytics. These businesses can manage the maintenance intervals as desired by the companies.

The Hadoop-based analytics find its application across broad spectrum of industries today. Applications like retail management use these for site selection, brand analysis, loyalty programs, market-based analysis along with sentiment analysis of the products. The financial service-providing organizations leverage Hadoop for fraud detection and risk assessment. Hadoop-based analytics are used by government agencies for applications that relate to law enforcement, public transportation, national security, health and public safety (Hortonworks 2013).

2.7.3 Apache SPARK

It is an open-source framework that is available for processing of big data. It was inititally developed in the AMP Lab at U.C. Berkeley in 2009 and later was later open sourced in 2010 as an Apache project. The concept behind Spark is to provide a memory abstraction which allows efficient sharing of data. The data is shared across different stages of a map-reduce job. It also provides in-memory data sharing. It provides a comprehensive, unified framework which can manage big data processing requirements for data sets that are from myriad sources. These sources could be real-time data streaming or batch processing, online Web-sourced data, etc. Hadoop cluster application is very fast and can execute up to a hundred times faster in memory and ten times faster when running on disc (Shanahan and Dai 2015). Deployment of Spark can be done in different ways. It can provide native bindings for Java, Python, Scala and R programming languages. It also supports streaming data, SQL, machine learning along with graph processing. Apache Spark provides the potential and power of big data along with support for real-time analytics to the business organizations.

Spark Ecosystem Apache Spark is an open source cluster computing system. It consists of libraries and framework ecosystems for advanced data analytics. Apache Spark is a powerful and easy to use tool and is more productive as compared to the MapReduce. It also provides in-memory, faster runtimes and support for distributed computing. Some other libraries in addition to the Spark Core API library are also part of the Spark Ecosystem. These are capable of providing advanced capabilities for big data analysis as shown in Fig. 2.7 (Hightower and Maalouli 2015; Penchikala 2015).

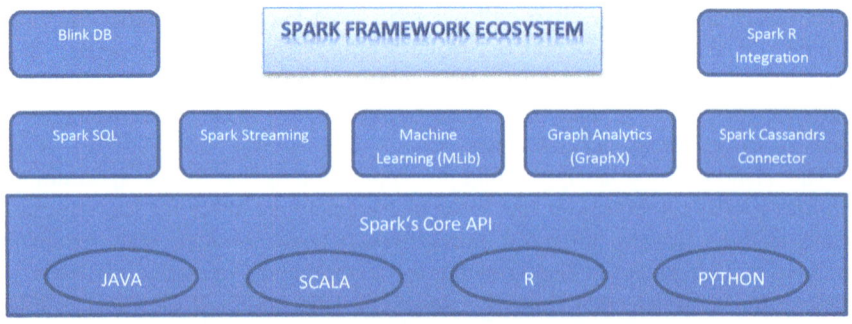

Fig. 2.7 Apache Spark Ecosystem (Hightower and Maalouli 2015; Penchikala 2015)

The base is the core engine on which the entire ecosystem is built. The API has support for Scala, Java, Python and R programming languages. Various libraries which provide additional computational power to spark are as follows:

- Spark Streaming can be used to process real-time streaming data which is based on the micro-batch style (i.e. splitting the input data into small batches) of computing and processing. To process the real-time data stream, DStream is used, which is a series of resilient distributed data sets (RDDs) (Zaharia et al. 2016).
- Spark data sets can be uncovered by Spark SQL which runs over the JDBC API. With the help of the traditional business intelligence along with the help of the visualization tools, Spark SQL allows running of SQL-like queries on big data. It also helps the users in data extraction from different formats (like Parquet, JSON or a database), transforming it and then finally exposing using to handle ad hoc queries.
- The machine learning library of Spark is the MLlib. It consists of both supervised and unsupervised machine learning algorithms, which include data classification, data clustering, linear and logistics regression, collaborative filtering, dimensionality reduction of data and optimization primitives("Big Data Processing with Apache Spark—Part 1: Introduction", n.d.).
- To compute graphs, the Spark GraphX component of Spark API is used. GraphX extends the Spark resilient distributed data set (RDD). It introduces the resilient distributed property graph which is a directed multigraph along with the properties that are associated with every edge and every vertex. GraphX also includes a collection of graph algorithms for simplifying graph analysis.

There are also adapters for integration with other products like Cassandra (Spark Cassandra Connector) and R (SparkR). The Cassandra connector allows Apache Spark to access data that is stored in a Cassandra database for data analysis.

2.8 Challenges of Big Data

With the growing size of Big data and the use of analytics, many challenges are uncovered. They are presented as under (Saravanakumar and Nandini 2017):

- **Size–Volume**
  New technologies have been proposed to facilitate the user and allow to store and query large data sets. However, the volume of data that is generated today is enormous; hence, new techniques with new algorithm along with new technology platform and ability to understand the data structure and business values is essential. To handle such challenges, "Data Scientists" with multidisciplinary expertise are required.

- **Acceptance of Big Data**
  It involves client motivation so as to acknowledge big data as a medium or channel for accepting and adopting new procedures and system. The acceptance of the same is time-consuming because to understand the big data and analytics is a tough task.
- **Understanding Analytics**
  Understanding the analytics to reduce the size and improve the business value is a major challenge. This happens as the objects that have to be modelled are of different nature than the contemporary. They are huge, complex and distributed. To handle this challenge, modelling and simulation techniques are needed which should be simple, robust, distributed and parallel computing.
- **Capturing Data**
  The data that constitutes the big data is of different types. They can be unstructured, semi-structured or unstructured. To capture such data for analysis is also quite challenging for the business organization.
- **Data Curation**
  In the big data era, data curation indicates processes and activities that are related to the organization along with the integration of data that is collected from a variety of sources. The data curation has become important as the software processes very high volume of complex data. It also includes data annotation, publication and presentation. Technically, data curation indicates the process of extracting of relevant information from large data set of interest.
- **Data Visualization**
  There is a vibrant problem of data visualization. As the big data is enormous, the users when access complex information and handle associated tasks, there is a vibrant difficulty faced by the users. In order to face these challenges, system software need to be used judiciously.
- **Performance and Scalability**
  The two main concerns and challenges of storage and processing enormous volume of big data systems are performance and scalability. To accomplish this, process analysis can be used so as to improve the performance and scalability.
- **Distributed Storage**
  Big data storage depends on distributed storages as the huge volume of data can only be stored and accessed in distributed platforms. These storages have to be handled technically and intelligently so that the data is available on the go. Handling the high volume and high velocity of big data is also a big challenge. Security of the data in motion also poses a great amount of trial.
- **Content Validation**
  Another major challenge that the data of the Internet face is the content validation. A large number of data sources like blogs, social networking sites, tweets, comments have information that is difficult to validate. Automated validation may be performed by using the machine learning algorithms to extract and validate the Web content.

2.9 Conclusion and Future Trends

In the past decade, the data exploded and became bigger and bigger ranging from terabytes to petabytes to exabytes. The business intelligence has revolutionized in past few years. Cloud technology has gained the maximum acceptance. Business houses rely on this structure for their data storage. However, the data can be stored in big reservoirs termed as data lakes. Unlike the data that the traditional databases use, the big data comprises unstructured, semi-structured and structured data that is generated from a large number and a variety of data sources. The big data is said to be at rest when it is stored in a stable structure, whereas the data in motion is when the data is in transit and has not reached the repository. The missionary data structure storages took a backseat and big data provided an actionable and insightful data presented with visualizations and interactive business dashboards.

Business intelligence was in full boom in year 2017. Trends that are present in the year 2017 will continue in 2018, but additional trends in analytics will be seen. The adopted strategies for analytics will be increasingly customizable. The question for business organizations would be somewhat like *"What is the best solution that is available?"* for business and what opportunities can be explored. The expected analytics and business intelligence trends for 2018 include (Lebied 2017) use of artificial intelligence for business intelligence; use of analytics tools—predictive and prescriptive; data quality management; the multicloud strategy deployment; data governance; natural language processing; security concerns; chief data officer —roles and responsibility embedded and collaborative business intelligence.

References

Al-Barznji, K., & Atanassov, A. (2017). Collaborative filtering techniques for generating recommendations on big data. In *International Conference Automatics and Informatics* (pp. 225–228). Sofia, Bulgaria.

Beaver, D. (2008). *10 billion photos*. Retrieved from https://www.facebook.com/notes/facebook-engineering/10-billion-photos/30695603919/.

Bertolucci, J. (2013). *How ancestry.com manages generations of big data—Information week.* Retrieved from https://www.informationweek.com/big-data/big-data-analytics/how-ancestrycom-manages-generations-of-big-data/d/d-id/1112975?.

Big Data Processing with Apache Spark—Part 1: Introduction. (n.d.). Retrieved from https://www.infoq.com/articles/apache-spark-introduction.

Brock, V. & Khan, H. U. (2017). Big data analytics: Does organizational factor matters impact technology acceptance? *Journal of Big Data, 4*(1). https://doi.org/10.1186/s40537-017-0081-8.

Business Intelligence—BI—Gartner IT glossary. (n.d.). Retrieved from https://www.gartner.com/it-glossary/business-intelligence-bi/.

Data mining versus data analysis and analytics—Fraud and fraud detection—Academic library—free online college e textbooks. (n.d.). Retrieved from https://ebrary.net/13380/business_finance/data_mining_versus_data_analysis_analytics.

Haines, S. (2014). *Introduction to HBase, the NoSQL database for Hadoop | Introduction to HBase | InformIT.* Retrieved from http://www.informit.com/articles/article.aspx?p=2253412.

Hightower, R., & Maalouli, F. (2015). *Introduction to big data analytics w/ Apache Spark Pt. 1—DZone Big Data*. Retrieved from https://dzone.com/articles/introduction-to-bigdata-analytics-with-apache-spar-6.

Hortonworks, A. (2013). *The business analyst's guide to Hadoop get ready, get set, and go: A three-step guide to implementing Hadoop-based analytics*. Retrieved from https://hortonworks.com/wp-content/uploads/2013/01/Alteryx-Hadoop-Whitepaper-Final1.pdf.

Jain, V. K. (2015). *Big Data and Hadoop*. Khanna Publishers. Retrieved from https://Books.Google.Co.In/Books?Id=I6nodqaaqbaj&Printsec=Frontcover&source=gbs_ge_summary_r&cad=0#v=onepage&q&f=false.

Lebied, M. (2017). *Top 10 analytics & business intelligence trends for 2018*. Retrieved from https://www.datapine.com/blog/business-intelligence-trends/.

Manyika, J., Chui, M., Brad, B., Bughin, J., Dobbs, R., Roxburgh, C., & Hung, A. (2011). *McKinsey Global Institute The McKinsey Global Institute*. Retrieved from https://www.mckinsey.com/ ~ /media/McKinsey/Business-Functions/McKinsey-Digital/Our-Insights/Big-data-The-next-frontier-for-innovation/MGI_big_data_exec_summary.ashx.

Meet Hadoop—Hadoop: The definitive guide (3rd ed) (n.d.). Retrieved from https://www.safaribooksonline.com/library/view/hadoop-the-definitive/9781449328917/ch01.html.

Miller, S. (2014). Collaborative approaches needed to close the big data skills gap. *Journal of Organization Design, 3*(1), 26. https://doi.org/10.7146/jod.9823.

Oussous, A., Benjelloun, F. Z., Ait Lahcen, A., & Belfkih, S. (2017). Big data technologies: A survey. *Journal of King Saud University—Computer and Information Sciences*. Retrieved from https://doi.org/10.1016/j.jksuci.2017.06.001.

Penchikala, S. (2015). *Big data processing with apache spark—Part 1: Introduction*. Retrieved from https://www.infoq.com/articles/apache-spark-introduction.

Philip Chen, C. L., & Zhang, C. Y. (2014). Data-intensive applications, challenges, techniques and technologies: A survey on big data. *Information Sciences, 275*, 314–347. https://doi.org/10.1016/j.ins.2014.01.015.

Saravanakumar, R., & Nandini, C. (2017). A survey on the concepts and challenges of big data: Beyond the hype. *Advances in Computational Sciences and Technology, 10*(5), 875–884. http://www.ripublication.com.

Shanahan, J. G., & Dai, L. (2015). Large scale distributed data science using apache spark. In *Proceedings of the 21th ACM SIGKDD International Conference on Knowledge Discovery and Data Mining—KDD '15* (pp. 2323–2324). New York, New York, USA: ACM Press. https://doi.org/10.1145/2783258.2789993.

Stubbs, E. (2014). In *Wiley Big Data Series: Big data, big innovation : Enabling competitive differentiation through business analytics*.

Vossen, G. (2014). Big data as the new enabler in business and other intelligence. *Vietnam Journal of Computer Science, 1*(1), 3–14. Retrieved from https://doi.org/10.1007/s40595-013-0001-6.

Want to make big bucks in stock market? Use big data analytics. Retrieved from https://analyticsindiamag.com/want-make-big-bucks-use-big-data-analytics/.

What is big data?—Gartner IT glossary. Retrieved from https://www.gartner.com/it-glossary/big-data.

Zaharia, M., Franklin, M. J., Ghodsi, A., Gonzalez, J., Shenker, S., Stoica, I., et al. (2016). Apache Spark: A unified engine for big data processing. *Communications of the ACM, 59*(11), 56–65. https://doi.org/10.1145/2934664.

Chapter 3
Application of Panel Quantile Regression and Gravity Models in Exploring the Determinants of Turkish Automotive Export Industry

Ibrahim Huseyni, Ali Kemal Çelik and Miraç Eren

Abstract This paper purposes to determine potential factors influencing the amount of Turkish automotive industry exports. For this purpose, the available data of 68 major trading partners of Turkey in terms of automotive industry exports were utilized for the sample period 2007–2015. Both panel quantile regression and the gravity model of trade approaches were considered to analyze the relevant data. The empirical findings of this paper revealed that the population and the distance variables were found as statistically significant for all quantiles, while the former has positive and the latter has negative signs as expected. Results also indicated that there was a statistically significant positive correlation between GDP per capita and the amount of Turkish automotive industry exports at 10 and 50% quantiles; however, it was not statistically significant at 90% quantile despite its positive sign. Among Turkey's exporter countries, being a EU member country dummy variable was found to have a statistically significant positive impact on the amount of automotive industry exports. Real exchange rate was not found as a significant determinant of the amount of automotive industry exports. In the lights of empirical evidence obtained from this study, several recommendations were made for Turkey's future international trade policies.

Keywords Automotive industry · Export · Quantile regression
Gravity model · Panel data

I. Huseyni
Şırnak University, Şırnak, Turkey
e-mail: ibrahim_huseyni@hotmail.com

A. K. Çelik (✉)
Ardahan University, Ardahan, Turkey
e-mail: alikemalcelik@ardahan.edu.tr

M. Eren
Ondokuz Mayıs University, Samsun, Turkey
e-mail: mirac.eren@omu.edu.tr

© Springer Nature Singapore Pte Ltd. 2019
H. Chahal et al. (eds.), *Understanding the Role of Business Analytics*,
https://doi.org/10.1007/978-981-13-1334-9_3

3.1   Introduction

Economic growth is widely regarded as one of the most important macroeconomic goals of every country to boost employment and general welfare. In that context, there exists a strong empirical evidence in the existing literature that export-led expanding markets have positively contributed to economic growth by providing more efficient use of convenient resources since the late of 1800s (Emery 1967; Syron and Walsh 1968; Michaely 1977; Balassa 1978; Tyler 1981). As well as expanding markets, total exports also provide an indirect impact on economic development by increasing foreign exchange entry significantly. Developing countries require to increase fixed capital formation in numbers to increase their levels of revenues and to converge to developed countries. However, developing countries necessitate to import several commodities as their levels of technology lack to produce fixed capital formation. Nevertheless, their current foreign exchange reserves numerically are one of the most significant obstacles to their total imports. At that point, the total exports of a country play a key role to overcome the corresponding import restriction. Recent research (Yaprakli 2007; Altıntaş and Çetintaş 2010; Doru and Ersungur 2014; Kaya and Hüseyni 2015; Hüseyni and Çakmak 2016) that concerns on the impact of exports on economic growth confirms that exports positively contribute to economic growth of a country.

Recent rapid developments particularly on transportation and communication industries have led many automotive companies to transfer some of their production into developing countries due to keep closeness to relevant market, to utilize from inexpensive labor force and to avoid relatively high environment taxes. The current policies of automotive companies have been appeared as an alternative foreign exchange source for developing countries and the magnitude of automotive production and exports in developing countries have significantly increased. Moreover, the transfer of operations in the automotive industry into developing countries stands for expanding total exports, accordingly an increase in the supply of foreign exchange and more efficient finance of the imports of investment goods. As a developing country, Turkey also takes advantage of the present policies of multinational automotive companies while Turkish automotive exports have experienced a significant growth trend particularly after post-1996 period when Customs Union Agreement has come into force. The corresponding trend has led the total exports of the Turkish automotive industry to become one of the leading industries in Turkish economy, and numerically, the Turkish automotive industry has been an indispensable industry for total Turkish production and exports after post-2000 period.

Before the period of 2000s, labor-intensive textile industry has provided the most crucial contribution to Turkish exports in numbers, whereas automotive industry comes into prominence at the present time. Figure 3.1 illustrates the total exports of automotive industry in Turkey since 1989. As seen in Fig. 3.1, there is an enormous increase of the Turkish automotive exports after the year 2000 and the total automotive exports have overwhelmingly reached to $18 billion in 2008 compared to $800 million in 1998. As mentioned above, Customs Union

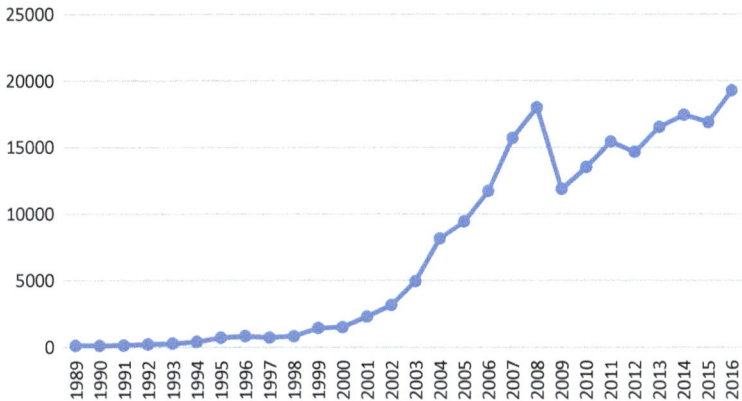

Fig. 3.1 Total automotive industry exports in Turkey between 1989 and 2016 (in $ million)

Agreement between Turkey and the European Union (EU) have significantly motivated the corresponding increase since the Agreement wholly eliminates certain tariffs on industrial goods. Not surprisingly, many multinational companies have carried out a remarkable amount of their automotive investments to the Turkish market and they have imported their products with inexpensive labor costs and environmental standards to specific regions including Europe, Middle and Near East. Thus, this circumstance stimulated the Turkish automotive exports to grow dramatically after post-2000 period. On the other hand, the negative impact of the 2008 Global Economic Crisis led the Turkish automotive industry exports to a remarkable decrease while the Turkish automotive industry exports have not been able to reach the previous favorable status before the crisis until 2015.

Figure 3.2 depicts a comparison between total exports and automotive industry exports of Turkey between 1989 and 2016 to better examine substantial changes relatively during the sample period. The blue line in Fig. 3.2 represents an index of the total amount of automotive industry exports while the claret red line represents an index of the total amount of exports for Turkey. In Fig. 3.2, an index was introduced by taking the numbers of 1989 year as the base and representing them as 100. As shown in Fig. 3.2, a more significant change trend has been experienced on the Turkish automotive industry exports than total exports in Turkey. Specifically, along with increases after 2000, the Turkish automotive industry exports have displayed a 140-time increase with respect to only a ten-time increase on total exports. In this sense, the impact of the 2008 Global Economic Crisis revisits, when both the Turkish automotive industry and total exports have significantly decreased in numbers in 2009. While the total exports have exhibited a strong and rapid recovery behavior after 2009, the Turkish automotive industry exports were not able to achieve such a short-term improvement numerically. One of the main reasons behind this situation may be explained as prolonged period under the negative influence of the 2008 Global Economic Crisis in the EU compared to rest of the world that led to the decline of supply. As the exports of the EU states

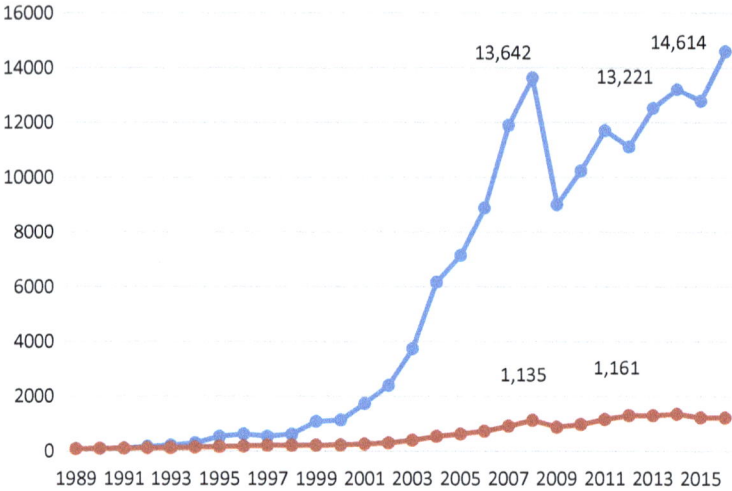

Fig. 3.2 Comparison between automotive industry and total exports of Turkey between 1989 and 2016

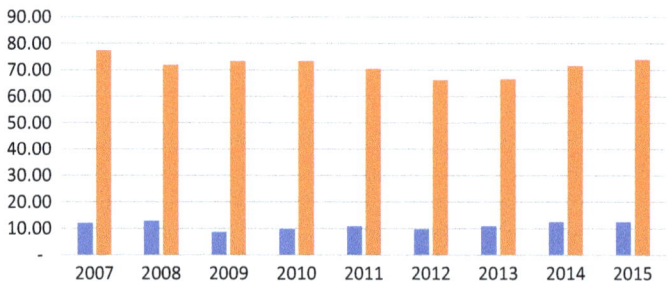

Fig. 3.3 Total exports of Turkey to the EU (in $ billion)

generate a majority of total exports of Turkey, the amount of Turkish automotive industry exports has been negatively affected.

Particularly, Fig. 3.3 illustrates exports of automotive products to the EU and the share of exports of automotive products to the EU on total exports of Turkey between 2007 and 2015. In Fig. 3.3, blue bars represent the total amount of exports to the EU member countries, while orange bars represent the share of the amount of automotive industry exports on the total amount of automotive industry exports. As seen in Fig. 3.3, the share of exports of automotive products to the EU on total exports varies between 70 and 80% during the sample period. This circumstance implies that Turkish automotive industry exports to the EU are highly dependent on the potential changes of automotive industry products supply in the EU.

3.2 Methodology

Quantile regression analysis provides a practical framework to examine how covariates make an impact on location, scale, and shape of the whole response distribution with an emphasis of classical least squares regression on the conditional mean (Koenker 2005). The generalization of conditional mean models is widely regarded as essential tools for the field of social sciences. Conditional mean models are able to explain a complete and parsimonious description of the association between the response distribution and the covariates and they provide maximum-likelihood and least-square estimators more useful in terms of calculation and interpretation (Hao and Naiman 2007).

Quantiles and quantile functions present a valuable information of summary measures when the distributions increasingly become less symmetrical and the quantile regression model specifies the condition quantile function. Theoretically, the pth quantile $Q^{(p)}$ of a cumulative distribution function F is accepted as the minimum set of values y that satisfies $F(y) \geq p$. In this context, as a function of p, the function $Q^{(p)}$ is defined as the quantile function of cumulative distribution function F (Hao and Naiman 2007).

The quantile regression model estimates the possible differential impact of a covariate on a variety of quantiles in the conditional distribution. A quantile regression model can be briefly defined in regard to the linear regression model as the following:

$$y_i = \beta_0^{(p)} + \beta_1^{(p)} x_i + \varepsilon_i^{(p)} \tag{1}$$

where $0 < p < 1$ denotes the proportion of the population having scores less than the quantile at p. Further, the pth condition quantile given x_i can be specified as

$$Q^{(p)}(y_i|x_i) = \beta_0^{(p)} + \beta_1^{(p)} x_i + Q^{(p)}(\varepsilon_i) \tag{2}$$

that can also be considered as the determination of the conditional pth quantile by quantile-specific parameters (i.e., $\beta_0^{(p)} + \beta_1^{(p)}$) and a specific value of covariate x_i. When the error terms ε_i are introduced, the equation

$$Q^{(p)}(y_i|x_i) = \beta_0^{(p)} + \beta_1^{(p)} x_i + Q^{(p)}(\varepsilon_i) \tag{3}$$

can be obtained (Hao and Naiman 2007).

Gravity models of international trade were introduced by Tinbergen (1962) and Pöyhönen (1963; Oguledo and Macphee 1994; Cheng and Wall 1999) and gravity model studies have been commonly utilized in explaining a wide variety of international and interregional association such as commuting, international trade, and labor migration (Cheng and Wall 1999). The gravity model is usually applied to international trade flows to state the size or the magnitude of trade flows between two countries (Oguledo and Macphee 1994). The traditional concept of the gravity equation suggests that GDP and GDP per capita in numbers and both preference factors such as common border and language and trade impediment such as distance are main explanatory variables of bilateral trade (Egger 2002). In other words, the distance is described as a function of several variables that can be considered as trade resistance factors (Metulini et al. 2018) The gravity equation that facilitate to explain bilateral trade flows across pairs of countries is defined as the following:

$$PX_{ij} = \psi_0 (Y_i)^{\psi_1} (Y_i/L_i)^{\psi_2} (Y_j)^{\psi_3} (Y_j/L_j)^{\psi_4} (D_{ij})^{\psi_5} (A_{ij})^{\psi_6} e_{ij} \tag{4}$$

In Eq. (1), PX_{ij} denotes the value of the flow from country i to country j in US dollars; $Y_i(Y_j)$ denotes the value of nominal GDP in $i(j)$ in US dollars; $L_i(L_j)$ denotes the total population in $i(j)$; D_{ij} denotes the distance from the economic center of i to the economic center of j; A_{ij} denotes any other factor(s) that has an impact on trade between i and j, and finally, e_{ij} denotes the error term with a log-normal distribution. The estimates of ψ_1, ψ_2, ψ_3, ψ_4 are representatively expected to be positive, whereas the estimates of ψ_5 are expected to have a negative sign (Bergstrand 1989).

Quantile regression analysis and the gravity models of trade applications take their respectable place in the existing literature. Dufrenet et al. (2010) explore the variation of the impact of trade openness on the growth rate of per capita income with the conditional distribution of growth using a quantile regression approach and they suggest a heterogeneous trade versus growth association for both short and the long run. Using the data on aggregate bilateral sales in 2008 for 93 economies, Baltagi and Egger (2016) find that trade costs show a differentiation across the quantiles of the conditional distribution of bilateral exports. Using a quantile regression approach, Trinh and Doan (2018) found that internationalization was positively correlated with several variables including the growth of employment, output, and labor productivity for Vietnamese enterprises. Özer (2014) investigates the determinants of the textile production exports of Turkey for the sample period 2007–2012 using a gravity of trade and quantile approaches and finds that there exists a statistically significant association between the total population and a demand increase. On the contrary, Tatlıcı and Kızıltan (2011) find that the population and the amount of Turkish exports, while the distance was negatively correlated. Martínez-San Román et al. (2016) find evidence on the positive association between trade integration and foreign direct investment activity for their selected countries. Very recently, Metulini et al. (2018) perform a spatial-filtering zero-inflated approach to estimate the gravity model of trade which they argue to be considered when the level of trade between countries is zero. (See Egger and Pfaffermayer 2016; Egger and Staub 2016; Spring and Grossmann 2016 for further successful applications of gravity models of trade).

The dependent variable of this study was selected as the total amount of automotive industry exports of Turkey to country i in year t. The data of the dependent variable were drawn from the United Nations' Comtrade database (United Nations 2018). The independent variables that may possibly influence Turkish automotive industry of Turkey to country i in year t were explained in detail as the following.

3.2.1 GDP Per Capita

GDP per capita was included as an independent variable in the estimated model. As automotive industry products are not inferior goods, an increase on the GDP per capita for country i, namely an increase of revenues for the corresponding country, will increase the export capacity that also leads to increase automotive demand. An increasing automotive demand will inherently increase the amount of automotive imports for country i. The sign of the GDP per capita is expected to be positive. The GDP per capita data of selected countries was drawn from the World Bank database (The World Bank 2018).

3.2.2 Population

The total population of exporter country i to Turkey is the second independent variable that was included in the estimated model. An increase on the population of a country will encourage to an increase on the amount of total demand by providing a potential increase on the amount of consumption. The increasing total demand will have a positive impact on automotive industry exports by influencing the amount of total imports in proportion as the ratio of import tendency. The populations of exporter countries were also drawn from the World Bank database (The World Bank 2018) and the population variable is expected to have a positive sign.

3.2.3 The Distance Between Exporter Country to the Capital of Turkey

The total distance among countries will boost transportation costs, while the foreign trade capacity will tend to decrease numerically. In fact, recent foreign trade data put forward that relatively higher amount of foreign trade among neighbor countries. As high volume goods will also lead to increase transportation costs, the distance and foreign trade are widely regarded as negatively correlated. Since many goods of automotive industry have relatively higher volume, the distance between capital cities of exporter countries to the capital city of Turkey (i.e., Ankara) was included in the estimated model as an independent variable.

3.2.4 The Real Exchange Rate

The real exchange rate is available for both countries; therefore, it should be carefully used in the estimated model. An increase on real exchange rate for any country i (i.e., appreciation of the currency) will encourage to increase the amount of automotive demand to Turkey. Nevertheless, if the real exchange rate has declined (i.e., depreciation of the currency) in the same period, the Turkish exporters will tend to sell their automotive industry products with higher prices. In that case, a consumer in country i will no longer use the increase on purchasing power by courtesy of the appreciation of the currency for imported automotive industry products from Turkey. This situation appears to avoid the increasing purchasing power by an increase on real exchange rate of country i to have a potentially positive impact on the amount of imports from Turkey. Moreover, if the decline on the real exchange rate is numerically less than an increase on the real exchange rate of country i, then the amount of imports of country i from Turkey may even decrease. In order to reflect the impact of actual changes of the real exchange rates for both countries, the real exchange rate of Turkey was divided into the foreign exchange rate of country i and the corresponding ratio was included in the final model being estimated. The data of the real exchange rates were drawn from multiple databases including the World Bank and Federal Reserve Economic Data (The World Bank 2018; Federal Reserve Bank of St. Louis 2018). The real exchange rate is expected to have a negative sign in the estimated model.

3.2.5 The Dummy Variable

As previously stated, the EU member countries were among the most important exporter countries of Turkey, especially after the Customs Union Agreement has come into force in 1996. When the prominent potential of the EU member countries were considered, being a EU member country is expected to be a relatively important determinant of the estimated model since Turkey will more likely to export to a EU member country than any other country in the rest of the world with respect to the recent total export statistics of Turkey. Therefore, a dummy variable was generated where 1 stands for being a EU member country and 0 stands for not being a EU member and was included in the estimated model. Hence, the marginal effect of being a EU member country on the demand of automotive industry products to Turkey would be examined. The dummy variable introduced to the model was expected to have a positive sign.

3.3 Empirical Evidence

The main objective of the present study is to examine the main determinants that may influence Turkey's vehicle (car, minibus, bus, van, and truck) exports to its major 68 trading partners over the period of 2007–2015. For this purpose, a gravity model and a panel data quantile regression approaches were performed calculated by the Bootstrap Method to obtain more consistent empirical results. The gravity

Table 3.1 Descriptive statistics of variables

Variables	Min.	1st Qu. 25% quantile	2nd Qu. 50% quantile	3rd Qu. 75% quantile	Max.	Mean	Skewness	Kurtosis
ln E_{it}	5.704	14.731	17.532	18.869	21.736	16.772	−0.560	−0.260
ln Y_{it}	6.016	8.611	9.624	10.639	11.674	9.499	−0.470	−0.660
ln P_{it}	12.650	15.470	16.220	17.630	21.040	16.460	−0.010	0.040
ln DIS_i	6.659	7.612	8.164	8.997	9.717	8.258	0.020	−0.950
$RERT_{it}$	0.600	0.870	0.920	0.990	1.240	0.923	0.160	1.310

models take an increasingly attention to explain exports and trade for panel data analyses in the existing literature. An econometric model was introduced to determine factors that may possibly have an impact on the automotive industry product exports of Turkey as the following:

$$\ln E_{it} = \beta_0 + \beta_1 \ln Y_{it} + \beta_2 \ln P_{it} + \beta_3 \ln DIS_i + \beta_4 RERT_{it} + \beta_5 D_{EU} + u_{it} \quad (5)$$

where E_{it} denotes the total automotive industry exports (in US dollars) from Turkey to the i country in year t; Y_{it} denotes the GDP per capita of importer country i in year t; P_{it} denotes the total population of the importer country i in year t; DIS_i denotes the geographical distance from Turkey to country i; $RERT_{it}$ denotes the ratio of Turkey's real exchange rate to real exchange rates of country i in year t; and finally, D_{EU} denotes the dummy variable of automotive industry exports to a country where stands for 1 if it is a member of the EU and 0 where it is not a member of the EU As seen in Eq. (1), some of the variables were measured in logarithmic terms. Table 3.1 presents the descriptive statistics of the variables used in the estimated model.

Table 3.1 indicates that different quantiles will describe different distribution tendency. For instance, when both mean and second quantile (i.e., median) values of selected variables are compared, and one can notice that the distribution of variables will substantially differ. Therefore, an ordinary least squares (OLS) regression approach may generate biased results (Santos Silva and Tenreyro 2006). As seen in Table 3.1, all variables are appeared to be skewed that confirms the use of quantile regression approach for the estimated model to determine potential factors that may have an impact on the automotive industry exports of Turkey. A set of data was collected from some databases related to the explanatory variables.

Table 3.2 summarizes the panel quantile regression analysis results to determine potential factors that may influence automotive industry exports of Turkey for the sample period 2007–2015. Due to the different levels of the GDP per capita variable, the impact of other independent variables on Turkey's automotive industry exports may also regard as different. Thus, this study adopts three different levels of the GDP per capita as three different levels of income and measures the direct

Table 3.2 Quantile regression results for the amount of Turkish automotive industry exports

	Quantiles		
Variables	10th quantile	50th quantile	90th quantile
ln Y_{it}	2.354 (0.022)*	2.545 (0.000)**	0.619 (0.287)
ln P_{it}	1.222 (0.000)**	1.144 (0.000)**	1.139 (0.000)**
ln DIS$_i$	−1.714 (0.008)**	−1.408 (0.000)**	−1.050 (0.000)**
RERT$_{it}$	0.375 (0.676)	0.820 (0.175)	−0.502 (0.298)
D_{EU}	1.170 (0.022)*	0.439 (0.004)**	0.206 (0.295)

Notes Values in parentheses are significancy probabilities
*Statistical significance at 5%
**Statistical significance at 1%

effects of selected independent variables on the automotive industry exports of Turkey to its major trading partners between 2007 and 2015.

Quantile regression analysis results in Table 3.2 present valuable information about the determinants of automotive industry exports of Turkey. In regard to high and low quantiles, the empirical evidence highlights that there exist significant differences for the impact of GDP per capita for selected countries on the conditional distribution of the amount of automotive industry exports to major trading partners. Particularly, the impact of GDP per capita on automotive industry exports was statistically significant and remarkable in the 10th quantile. For instance, several countries including Equatorial Guinea, Trinidad and Tobago, Dominican Republic, Paraguay, Costa Rica, Zambia, the Philippines, Iceland, Uganda, and Uruguay purchase automotive products with relatively low commercial values. On the other hand, the impact of GDP per capita on the automotive industry exports was not statistically significant for higher quantiles. In lower levels of quantiles, one of the most important barriers on car ownership can be principally considered as comparatively low income. Certainly, when individuals' income increases, their automotive demand also increases which will make a positive impact on Turkish automotive industry exports.

In higher quantiles, where some high-income countries including France, the UK, Germany, Italy, Russian Federation, Spain, Belgium, and the USA are appeared, tastes and preferences take the place of income level to be an important determinants of automotive industry demand. Specifically, the automotive industry demand cannot be simultaneously increased with a potential increase on the amount of income since the income level that will increase automotive demand is eliminated in a sense. Moreover, a potential increase on the amount of income in the relevant quantile with high-income cases may reversely lead an increase on the consumption of luxury automotive products which are not exported, in Turkey. In this circumstance, an exclusion of middle-segment products may even expected for Turkey's further exports. However, the amount of GDP per capita of countries in high quantiles shows a decreasing tendency in the long run during the selected sample period. For that reason, the positive correlation between the amount of GDP per capita and Turkish automotive industry exports was not associated for all quantiles being observed. In this sense, one can argue that there exists an exact heterogeneity for the impact of different values of GDP per capita on the amount of automotive industry exports.

The population variable used in the model was found to have a statistically significant positive impact on the amount of Turkish automotive industry exports. In other words, a respectable increase on total population numerically leads to an increase on total consumption, and thus, exporter countries tend to increase their imports from Turkey including automotive industry products. Results revealed that there was a statistically significant correlation between the distance variable and the amount of automotive industry exports for all quantiles as well. As expected, the distance variable was found to have a negative sign that implies the amount of Turkish automotive industry exports decreases relatively when the distance from the capital of Turkey to the exporter country increases. The dummy variable, namely, being a EU member country was found as a statistically significant variable with a positive impact on the amount of Turkish automotive exports in the first two quantiles. This evidence was actually expected; however, being a EU member country was unexpectedly not found to have a statistically significant impact on the amount of Turkish automotive exports in higher quantiles despite its expected positive sign.

The real exchange rate was expected to have a negative sign before fitting the panel quantile regression model; however, it was not found as statistically significant. Accordingly, one can suggest that the real exchange rate may not essentially be an important determinant of Turkish automotive exports during the sample period. One explanation for this outcome may be comparatively high amounts of imported output numerically in the automotive industry. Though a real exchange rate appraisal contributes to competition power of exporter enterprises in the automotive industry, its actual impact on the amount of Turkish automotive exports is somehow deteriorated due to higher numbers of Turkish imports comparatively.

3.4 Main Conclusions and Policy Implications

Economic growth is an essential goal of every country to be accomplished because a sustainable economic growth has a crucial role on improving employment and welfare indicators. Recent debates since the late of eighteenth century suggest that the amount of exports is one of the most important determinants of economic growth when its impact on expanding markets and enabling the efficient use of resources. The amount of exports can especially provide developing countries an opportunity to converge to developed countries through increasing fixed capital investments numerically with foreign exchange entry.

Recent rapid developments on communication and transportation industries and simultaneous remarkable influence of globalization have led multinational companies to carry their production operations to developing countries. Particularly, multinational companies take advantage of cheaper labor force and avoid strict environmental standards in developing country markets. Thus, developing countries are able to have increasing production and exporting opportunities. In that sense, automotive industry can be considered as the leading industry in terms of its high potential of exporting and the production operations in the automotive industry have been significantly increased. After the 1996 Custom Unions Agreement with the EU, many multinational automotive companies have shifted a respectable number of productions to the Turkish market by courtesy of eliminating customs tariffs. Thus, the amount of Turkish automotive industry exports has significantly improved since 2000s.

When the importance of automotive industry exports for the Turkish economy is considered, the determination of factors influencing the amount of Turkish automotive industry exports gives valuable information for future foreign trade policies. This study aimed at determining factors affecting the amount of Turkish exports to 68 major trading partners using panel quantile regression and gravity approaches instead of OLS estimators frequently performed in the existing literature. The empirical evidence obtained from the estimation results revealed that the population of importer country and the amount of per capita income were found as positively correlated with the amount of Turkish automotive exports. Additionally, when the distance between importer country and the capital of Turkey increases, the amount of Turkish automotive industry exports was more likely to have a decreasing behavior. As expected, exporting a EU member country was found to have a statistically significant increasing impact on the amount of automotive industry exports. The estimation results also indicated that the real exchange rate was not a statistically significant determinant of the amount of automotive industry exports during the sample period. Turkey cannot exactly succeed to use the competitive advantage of the possible declines on real exchange rates due to higher costs of imports in the automotive industry. Consequently, potential increases on the population and revenues of importer countries of Turkey were found to be significantly effective on Turkish automotive industry exports. The amount of automotive industry exports may be increased in the coming years, after the devastating impacts of the 2008 Global Economic Crisis in the EU completely disappear. Further, foreign trade policies in Turkey may also concentrate on decreasing the importing costs of the automotive industry to take the advantage of a decrease on real exchange rates.

References

Altıntaş, H., & Çetintaş, H. (2010). Türkiye'de ekonomik büyüme, beşeri sermaye ve ihracat arasındaki ilişkilerin ekonometrik analizi: 1970–2007 (in Turkish). Erciyes Üniversitesi İktisadi ve İdari Bilimler Fakültesi Dergisi. 36, 33–56. http://dergipark.gov.tr/download/article-file/66628.

Balassa, B. (1978). Exports and economic growth: Further evidence. *Journal of Development Economics, 5*(2), 181–189. https://doi.org/10.1016/0304-3878(78)90006-8.

Baltagi, H., & Egger, P. (2016). Estimation of structural gravity quantile regression. *Empirical Economics, 50*, 5–15. https://doi.org/10.1007/s00181-015-0956-5.

Bergstrand, J. H. (1989). The generalized gravity equation, monopolistic competition, and the factor-proportions theory in international trade. *Review of Economics and Statistics, 71*(1), 143–153. https://doi.org/10.2307/1928061.

Cheng, I. H., & Wall, H. J. (1999). Controlling for heterogeneity in gravity models of trade and integration. *Federal Reserve Bank St, 87*(1), 49–63.

Dufrenet, G., Mignon, V., & Tsangarides, C. (2010). The trade-growth nexus in the developing countries: A quantile regression approach. *Rev World Econ, 146*, 731–761. https://doi.org/10. 1007/s10290-010-0067-5.

Egger, P. (2002). An econometric view on the estimation of gravity models and the calculation of trade potentials. *The World Economy, 25*(2), 297–312. https://doi.org/10.1111/1467-9701. 00432.

Egger, P., & Pfaffermayr, M. (2016). A generalized spatial error components model for gravity equations. *Empirical Economics, 50*, 177–195. https://doi.org/10.1007/s00181-015-0980-5.

Egger, P., & Staub, K. E. (2016). GLM estimation of trade gravity models with fixed effects. *Empirical Economics, 50*, 137–175. https://doi.org/10.1007/s00181-015-0935-x.

Emery, R. F. (1967). The relation of exports and economic growth. *Kyklos, 20*(4), 470–486. https://doi.org/10.1111/j.1467-6435.1967.tb00859.x.

Ersungur, S. M., & Doru, Ö. (2014). Türkiye'de dış ticaret ve ekonomik kalkınma ilişkisinin ekonometrik analizi (1980–2010) (in Turkish). *Atatürk Üniversitesi İktisadi ve İdari Bilimler Dergisi, 28*(3), 225–240. http://dergipark.gov.tr/atauniiibd/issue/2714/36026.

Federal Reserve Bank of St. Louis (2018) Federal Reserve Economic Data. https://fred.stlouisfed.org/.

Hao, L., & Naiman, D. Q. (2007). *Quantile regression*. London: Sage Publications.

Hüseyni, İ., & Çakmak, E (2016) Türkiye'de ihracat ve ekonomi büyüme ilişkisi: Eş- bütünleşme and nedensellik ilişkisi (in Turkish). *Atatürk Üniversitesi İktisadi ve İdari Bilimler Dergisi, 30* (4), 831–844. http://e-dergi.atauni.edu.tr/atauniiibd/article/view/5000185805/5000174888.

Kaya, V., & Hüseyni, İ. (2015) İhracatın sektörel yapısı ve ülkelere dağılımının ekonomik büyüme üzerindeki etkileri: Türkiye örneği. *Atatürk Üniversitesi İktisadi ve İdari Bilimler Dergisi, 29* (4), 749–773. http://dergipark.gov.tr/download/article-file/30571.

Koenker, R. (2005). *Quantile regression*. Cambridge: Cambridge University Press.

Martínez-San Román, V., Bengoa, M., & Sánchez-Robles, B. (2016). Foreign direct investment, trade integration and the home bias: Evidence from the European Union. *Empirical Economics, 50*, 197–229. https://doi.org/10.1007/s00181-015-0942-y.

Metulini, R., Patuelli, R., & Griffith, D. A. (2018). A spatial-filtering zero-inflated approach to the estimation of the gravity model of trade. *Econometrics, 6*(1), 1–15. https://doi.org/10.3390/ econometrics6010009.

Michaely, M. (1977). Exports and growth: An empirical investigation. *Journal of Development Economics, 4*(1), 49–53. https://doi.org/10.1016/0304-3878(77)90006-2.

Oguledo, V. I., & Macphee, C. R. (1994). Gravity models: A reformulation and an application to discriminatory trade arrangements. *Applied Economics, 26*(2), 107–120. https://doi.org/10. 1080/00036849400000066.

Özer, O. O. (2014). Türkiye'nin tekstil ürünleri ihracatının belirleyicileri: Çekim modeli yaklaşımı (in Turkish). *Tekstil ve Konfeksiyon, 24*(3), 252–258. http://dergipark.gov.tr/download/article-file/218239.

Pöyhönen, P. (1963). A tentative model for the volume of trade between countries. *Weltwirtsch Arch, 90*(1), 93–100. https://www.jstor.org/stable/40436776.

Santos Silva, J. M. C., & Tenreyro, S. (2006). The log of gravity. *Review of Economics and Statistics, 88*, 641–658. https://doi.org/10.1162/rest.88.4.641.

Spring, E., & Grossmann, V. (2016). Does bilateral trust across countries really affect international trade and factor mobility? *Empirical Economics, 50*, 103–136. https://doi.org/10.1007/s00181-015-0915-1.

Syron, R. F., & Walsh, B. M. (1968). The relation of exports and economic growth. *Kyklos, 21*(3), 541–545. https://doi.org/10.1111/j.1467-6435.1968.tb00131.x.

Tatlıcı, Ö., & Kızıltan, A. (2011). Çekim modeli: Türkiye'nin ihracatı üzerine bir uygulama (in Turkish). *Atatürk Üniversitesi İktisadi ve İdari Bilimler Dergisi, 25*, 287–299. http://dergipark. gov.tr/download/article-file/30446.

The World Bank. (2018). *Microdata library*. Washington, DC. Retrieved fromhttp://microdata. worldbank.org/index.php/home

Tinbergen, J. (1962). *Shaping the world economy: Suggestions for an international economic policy*. New York: The Twentieth Century Fund.

Trinh, L. G., & Doan, H. T. T. (2018). Internationalization and the growth of Vietnamese micro, small, and medium sized enterprises: Evidence from panel quantile regressions. *Journal of Asian Economics, 55*, 71–83. https://doi.org/10.1016/j.asieco.2018.01.002.

Tyler, W.G. (1981). Growth and export expansion in developing countries: Some empirical evidence. *Journal of Development Economics, 9*(1), 121–130. https://doi.org/10.1016/0304-3878(81)90007-9

United Nations. (2018). United Nations Comtrade International Trade Statistics Database. https://comtrade.un.org/.

Yapraklı, S. (2007). İhracat ile ekonomik büyüme arasındaki nedensellik: Türkiye üzerine ekonometrik bir analiz. *METU Studies in Development, 34*(1), 97–112. Retrieved from http://www2.feas.metu.edu.tr/metusd/ojs/index.php/metusd/article/view/153.

World Bank. World Bank open data. https://data.worldbank.org/.

Chapter 4
Impact of Macroeconomic and Bank-Specific Indicators on Net Interest Margin: An Empirical Analysis

Arif Ahmad Wani, S. M. Imamul Haque and Shahid Hamid Raina

Abstract The purpose of this paper is to identify the indicators from macroeconomic and bank environment, which tend to affect earning capacity (quantified by net interest margin) of public sector banks (PSBs) of India. The paper also quests to explore the possible linkages between the indicators under the purview of this paper. The financial statements, financial notes, and annual reports of the sample banks, publications from Government of India, Reserve Bank of India, and World Bank have been accessed to get the data regarding the variables under the study. The classical multiple regression analysis has been employed with diagnostic tests to derive concrete inferences from the data. The empirical evidences illuminated the positive correlation of gross domestic product (GDP), inflation, lending interest rate (LIR), and capital to risk-weighted assets ratio (CRAR) with the net interest margin (NIM) of sample banks, while as non-performing loans (NPLS) established an indirect relationship. The study established that favorable macroeconomic environment proves to be a main driver for encouraging net interest margin (NIM) with a prudent control over CRAR along with NPLs on the part of sample banks. The study suggested installing latest advances and practices of risk management especially on the credit front, which will also help the banks to utilize excessive capital rather than accumulating it unnecessarily. It is also suggested for the PSBs to merge for better consolidation, allocation of funds, and better investment prospects.

Keywords Net interest margin (NIM) · Indian Public sector banks (PSBs)
Economic growth · Inflation · Non-performing loans (NPLs)

A. A. Wani · S. M.I. Haque
Department of Commerce, Aligarh Muslim University, Aligarh 202002, UP, India
e-mail: aarifibniahmad@gmail.com

S. M.I. Haque
e-mail: imamul_haque62@yahoo.com

S. H. Raina (✉)
Department of Economics, Central University of Jammu, Jammu 181143, India
e-mail: rainashahid7@gmail.com

© Springer Nature Singapore Pte Ltd. 2019
H. Chahal et al. (eds.), *Understanding the Role of Business Analytics*,
https://doi.org/10.1007/978-981-13-1334-9_4

4.1 Introduction

Banking sector is considered to be the backbone for the growth of any economy. The economic development of a nation depends on the financial system it has instituted (King and Levine 1993; Rousseau and Wachtel 1998). Therefore, it is imperative to appraise the different components of financial system in order to invigorate the economy. Banking sector as one of the constituents of financial system needs greater alertness for it acts as a precursor to financial stability (Mishra et al. 2013). Moreover, more vibrant and stronger the banking sector, stronger and less fragile is its financial stability. In India, banking sector has witnessed major transitions and survived successfully since its inception from traditional banking practices, nationalization at different points of time to privatization with a number of private players both from within and outside the country, offering a diverse set of competitive services (Goyal and Joshi 2012; Haque and Wani 2015). Inter alia major transitions, economic reforms in 1991 unlocked new pinnacles of development but parallel to this posed new challenges and strains before the Indian banking sector. The major changes are spread across a spectrum of activities ranging from credit expansion, enhanced profitability and productivity almost at par with developed economies, lower incidence of non-performing loans (NPLs), financial inclusion, development in supervisory and regulatory insight to prudential and instrumental ways of advancing in terms of size, assets, profitability, and therefore, efficiency. Despite this progression, Indian banking industry is vulnerable to economic shocks and financial risks. The most prominent financial shocks across the globe like subprime of 2008 in USA, debt crisis in different economies including India, and euro zone crisis have exploded the confidence of financially vibrant economies and created an uncertain environment for the world economy. These economic downturns have raised different question marks about the survival and growth of economies throughout the globe. After these crises, the financial institutions are being subjected to stronger regulatory framework consistent with the international standards. Thus, to ensure financial stability of an economy, the lessons taught by these dynamite events were translated into regulatory policies, which were later strongly recommended to be instituted for prudence and conservatism. However, amidst all this pandemonium and turmoil, Indian banking sector has been among the few to recoup resilience. Notably, the momentum of development for the Indian banking industry has been gilt-edged over the past decade. It is evident from the higher pace of credit expansion with a slight control over NPLs, expanding profitability with minimum costs and focus on increasing the banking ambit through the schemes like financial inclusion. Such competencies have contributed to making Indian banking more vibrant and stronger. Indian banks like the developed economies have learnt to revamp their growth approach and re-evaluate the prospects to keep the economy rolling without interruptions. This is evident from the fact that Indian banks are striving to withstand against the competition from global banks as technological innovation has compelled the banks to rethink and revisit their policies and strategies (Goyal and Joshi 2012).

Financial intermediation offered by the banking sector supports economic growth by converting deposits into productive investments. There exists a strong connection between financial intermediation and economic growth as the funding costs have a significant impact on the investment level, capital allocation, and thus on growth potential and the direction of an economic activity (Claeys and Vander 2008; Kasman et al. 2010; Maudos and De Guevarra 2004). Financial intermediation also affects the profitability of the banking sector and therefore its stability and ability to support the real economy (García-Herrero et al. 2009). Indian banking sector is one of the cornerstones for financial stability and economic development of the country.

Therefore, it is quite imperative to examine the viability and sustenance of Indian banks and contribute to the literature by identifying and empirically investigating the impact of different indicators on the net interest margin of banks. This paper seeks to investigate the determinants, their linkages and impact on net interest margin of banks. To pursue the objectives, the study takes into account the public sector banks of Indian industry in order to turn up with potential explanations and policy implications which will help in obviating the undesirable practices.

4.2 Conceptual Framework

4.2.1 Net Interest Margin (NIM)

Commercial banks run business with a public role which includes money supply, payments system, insured deposits, and more (Gup and Kolari 2006). Banks play an indispensable role in modern economies by transferring the funds from savers to borrowers by incurring the costs of financial intermediation. The cost of financial intermediation is a prominent indicator of total financing costs. The primary source of income for commercial banks is the interest income. Banks maintain assets and liabilities of different magnitude and maturities and subsequently charge different types of interest rates for each category. Therefore, there is no fixed way of determining what they charge and pay. One of the best and widely used indicators to measure the difference is net interest margin (NIM), which is arrived at by dividing net interest income (NII) by the total earning assets of the bank. NIM is seen as an efficiency ratio as it best explains the employment of earning assets.

Research witnesses that the cost of financial intermediation has important consequences for the economy and banks. Thus, NIM has different repercussions for economies and banks. For the economy, consistently high NIMs might be considered as symptomatic for a spectrum of systemic problems like lack of competition, unsoundness of banks, higher operating costs due to low operating efficiency, perceived financial risks, scale of diseconomies, unfavorable environment, and presence of various regulatory obstructions, which in a synergetic manner distort the financial market activities. Such distortions due to higher NIMs further signal low efficiency of banks; higher costs and inefficient control of operating expenses, which together negatively affect the financial market development and get reflected in the sluggish movement in the economic growth. On the contrary, lower NIMs are treated as good indicators of financial market development, increased investments, and thus, higher economic growth. In juxtaposition to these influences, higher NIM has positive repercussions for the profitability and capital of banks and vice versa. At the same time, possible reason for lower NIMs may be thought that banks may be deliberately foregoing their NIM by availing more income through the delivery of non-interest bearing services like fees, commissions (Kasman et al. 2010). Therefore, NIM as a measure of income is seen differently through economic and banking perspectives.

4.2.2   Economic Growth

Indian economy has weathered many challenges successfully in recent times and is currently placed on the back of strong policies and a whiff of new optimism. With the dawn of reforms, the economy faced testing times with issues like lower economic growth, escalated levels of inflation, widening current account deficit, and interest rate volatility. Despite the moderate progression in gross domestic product (GDP) growth rate from 4.5% in FY13 to 4.7% in FY14, Indian macroeconomic ambience has proved to be moving less vigorously. At the macrolevel, this lackluster behavior can be attributed to a gamut of reasons like sluggish industrial growth, supply-side constraints, restrained demand conditions, inflation, and decelerated growth in services sector. All these factors synergistically leave a profound impact and pose greater challenges to the banks operating in India. At the bank level, rising NPAs, obsolete quality of assets, risk aversion behavior of banks, regulatory obstructions followed by reduced credit, and deposit base led to declining interest margins and profitability. This series of events eventually contributed to the lagging of economic growth. Among all the issues, the two main challenges having relevance and showing the relationship between earning potential of PSBs and economic growth are financing challenge and the banking challenge (Economic Survey of India 2014–2015).

PSBs were identified as one of the main reasons behind the sluggish performance of Indian economy. PSBs financed a significant portion of infrastructure in the country, but the deterioration of their balance sheets holds back such private investment. Such a situation can be observed from their stressed financial statements, which clearly indicate that PSBs alone account for over 12% NPAs. Therefore, this balance sheet syndrome decreases their ability of lending and credit extending potential to the private sector, which eventually gets reflected in the slowdown of the economy.

Indian banking is also presumed to be crippled by regulatory policies reflecting double financial repression which impedes competition in the sector. The double financial repression reduces the earnings of both savers and banks by misallocating capital to investors. It is evident from the Statutory Liquidity Ratio (SLR) (holding Govt. securities requirements) and priority sector lending (deployment of funds in less efficient ways). Further, there appears a significant variation among PSBs in terms of performance measured by prudence and profitability. It is also quite fla-grant to notice that the best PSB records its performance well below the level of an average private sector bank. For this performance, it is believed that indulging in social obligations placed PSBs at a competitive disadvantage. Financial repression has also risen since 2007, when escalated inflation has resulted in negative real interest rates and reduction in household savings. Therefore, to deal with such problems, a policy of 4Ds has been recommended which comprise of deregulation, differentiate, diversify, and disinter. The policy is pragmatic in the sense if financial repression on the liability side of the balance sheet is overcome, there will be fall in inflation, thus relaxing the asset-side financial repression. Eventually, SLR requirements will be eased and priority sector lending norms will be revisited to bring the liquidity back to the banks for further credit creation and delivery of other services. This will help them to gear up both NII and non-interest income and thus, contribute to economic growth and strengthen the economic development.

4.2.3 Inflation

Economic growth and inflation are often used to demonstrate economic stability and monetary or price stability. Inflation as a concomitant element along with other indicators decides the direction of economic growth, its linkage with other macro- and microlevel indicators spell out different implications. Research witnesses its medium- to long-term association with the financial stability and economic growth. On the contrary, escalated inflation or in other words price instability adversely affects financial stability (Dhal et al. 2011). In the common parlance, persistent inflation seems to entrench uncertainty over prices, which in turn negatively affects investment and consumption patterns in an economy. In the Indian context, infla-tion has marked a remarkable improvement which is evident by the Consumer Price Index (CPI) inflation of 4.4% in November 2014 against 11.2% in the yesteryear (Financial Stability Report 2014–2015). As a result, Reserve Bank of India (RBI) reduced the repo rate by 25 basis points (to 7.75%) and SLR by 50 basis points (to 21.5%) in the past year. Along with other measures to tackle the inflationary pressure in an economy, the central bank tightens the monetary policy. As a result, banks increase the lending rates to improve their interest margins.

4.2.4 Lending Interest Rates (LIR)

Lending interest rate is the bank rate that usually meets the short- and medium-term financing needs of the private sector. This rate is normally differentiated according to creditworthiness of borrowers and objectives of financing. As a general rule, higher the lending rate, higher will be the margins. But the literature has witnessed that interest rates adopt dynamic behaviors in the short, medium, and long terms. In the short run, higher interest rates contribute to the declining interest margins; while as in medium to long term, they propel the interest margins high and higher (Busch and Memmel 2017). Laconically, such an uncertain and unpredictable movement in the interest rates exerts pressure on the earnings of banking sector and thus, infuses interest rate risk. During the inflationary pressures, banks hike their interest rate to make the economy withstand against inflation as well as to boost their interest earnings (Poghosyan 2013).

Moreover, the financial system of a country is one of the most important sources of financing economic decisions related to consumption and investment, capital accumulation and technological innovations, aimed at medium-term productivity growth to more dynamic and sustainable rates of economic growth. Accordingly, the price of financing through bank loans (i.e., lending rates) and the efficiency of the banking system (as measured by interest rate spreads) are essential for the allocation of additional financial potential in the economy, and thus for the acceleration or sustainability of economic growth (Georgievska et al. 2011).

Interest rate risk in a banking sector is important for a variety of reasons. Firstly, fluctuations in interest rates tend to manifest undeterminable jumps rather than gradual jumps especially when interest rates are used to wipe the excess liquidity from the economy as well as to meet the monetary objectives of the government. Secondly, complying with the international standards of risk-adjusted capital does not encompass interest rate risk, which therefore requires continuous quantification, control, and thus greater alertness.

4.2.5   Capital to Risk-Weighted Assets Ratio (CRAR)

Capital is the blood of banking whose purpose is to ensure that banks can sustain unexpected losses of the assets they hold while still honoring withdrawals and other essential obligations. Capital adequacy is a measure of creditworthiness of the banks, which is a result of combination of factors like regulation, market pressures, and business strategy of the bank. Such a measure prevents banks from accepting risks more than their appetite and ensuring stability in the banking sector (Claeys and Vander 2008). Thus, lower ratio of capital adequacy signals a relatively risky position which may result in a negative coefficient (Berger 1995). On the other hand, higher ratio of capital represents prudent lending by banks, infers ability to cut down funding costs (Altunbaş et al. 1997), helps in borrowing less, serves as a cushion against non-performing loans, and finally increases the expected earnings by lowering the costs of financial distress including bankruptcy (Berger 1995). In the Indian context, capital adequacy will start becoming a big issue for the commercial banks in India, as they start gearing for growth and becoming compliant to

Basel III guidelines, implemented in India effective from April 01, 2013, which will be fully implemented in a phased manner by March 31, 2019.

The CRAR as per Basel II at the end of March, 2014 was recorded at a comfortable level of 13.02%, which declined to 12.75% in September 2014. Though satisfying the regulatory requirement for CRAR of 9%, capital positions suffered a decline due to capital deterioration by PSBs which need to be infused with more capital to make their operations productive (Financial Stability Report 2014–2015).

4.2.6 Non-Performing Loans (NPL)

Asset quality is an important indicator to assess health of the banks. With regard to PSBs, they have suffered significant deterioration which is verified by the statistics of gross NPAs of 4.5% in September 2014 against 4.1% in March 2014. The sectors which solely held 54% of total stressed advances of PSBs as on June 2014 included infrastructure, iron and steel, textiles, mining, and aviation with 17.5% alone with the infrastructure (Economic Survey 2014–2015).

In pursuance of problem loans, RBI has taken a number of measures to withstand against this problem. In the month of June, 2014, it introduced guidelines on '*Early Recognition of Financial Distress,*' '*Prompt Steps for Resolution & Fair Recovery for Lenders,*' and '*Framework for Revitalizing Distressed Assets in the Economy.*' Subject to the implementation of these measures, asset quality of PSBs is expected to improve in the near future. In the event of failure of these guidelines, banks could resort to the '*Debt Recovery Tribunals*' or seek legal assistance by way of '*SARFAESI Act, 2002*' or may sell their NPAs to Asset Reconstruction Companies (ARCs), other banks, or non-banking financial companies (NBFCs) having requisite skills of resolving non-performing assets (NPAs) smoothly and efficiently (RBI Bulletin, March 2015). Thus, NPAs are problematic for all commercial banks including PSBs as they primarily depend on interest payments for income.

4.3 Review of Literature

This section not only surveys what past studies have revealed but also appraises, encapsulates, and compares various scholarly works. There is ample of literature which verifies and empirically investigates the linkage between the variables under study.

With regard to relationship between capital adequacy and bank profitability, a positive relation was verified by Bourke (1989). The underlying hypotheses for this relationship put forth were bankruptcy and signaling hypothesis, where the former believes offsetting of such costs with the capital at disposal and latter conveys the conservative approach of banks by ignoring the potential investment opportunities.

The hypothesis of bankruptcy costs was further supported by Berger (1995) with a potential explanation that higher the expected chances of insolvency, capital adequacy ratio also increases to lower the bankruptcy costs and thus, combat the chances of failure. These evidences are also supported by Angbazo (1997), Haslem (1969), and Olalekan and Adeyinka (2013).

With respect to GDP and inflation, they are expected to affect profitability according to the economic conditions, viz. they bring about positive changes in an economy where financial markets are finely developed and influence an economy negatively with developing financial markets (Alexiou and Sofoklis 2009). A negative correlation between inflation and profitability of banks were observed by Guru et al. (2002), whereas positive association between the two was reported by Tan and Floros (2012). Banks earn higher margins through increased lending rates enforced as a result of anticipated high inflation, while as banks profitability suffers when inflation is unanticipated (Perry 1992). The same results were revealed by Demirgüç-Kunt and Huizinga (1999) for developing countries, and also they stated positive association between GDP and profitability of banks. In another study by Hoggarth et al. (2001), inflationary pressures were found to infuse complexities in the contemplation of loan processing. The association between inflation and profitability was reported subjectively like Jiang et al. (2003) for Hong Kong and Guru et al. (2002) for Malaysia observed positive relationship while as Demirgüç-Kunt and Huizinga (1999) for developing countries and Abreu and Mendes (2001) concluded with a negative relation between inflation and profitability. Positive correlation of GDP and inflation with the financial performance of banks was observed by Fadzlan and Kahazanah (2009). Moreover, impact of different macroeconomic variables like inflation and GDP on NIM was reported subjectively by Kasman et al. (2010), Beck and Hesse (2009), Horváth (2009), Claeys and Vander (2008), and Brock and Suarez (2000) in their studies. In another study conducted by Kanwal and Nadeem (2013), any noticeable contribution by macroeconomic variables like inflation, GDP, and real interest rate toward earnings of the banks were not observed. Instead, they recommended concentrating on other variables with prime focus on internal factors of banks.

Similarly, studies like Maudos and De Guevara (2004) and Angbazo (1997) have verified a significant positive effect of credit risk on net interest margin. Maudos and De Guevara (2004), Brock and Suarez (2000), and Saunders and Schumacher (2000) concluded that interest volatility has been found to bear a significant and positive effect on interest margin in different countries. Monetary policy also affects the profitability of banks indirectly through the revision of interest rates under different circumstances (Khan and Sattar 2014).

Undoubtedly, researchers have extensively enquired into the relationship between profitability and macro- as well as microlevel indicators. The almost visibly common effort can be traced from their studies is that profitability has been quantified by measures other than NIM. Consequently, there appeared paucity of literature which strikes a relationship taking NIM as a profitability measure and examining its relevance with macro- and microlevel indicators in terms of both direction and magnitude. To fill the gap, the researchers believe this study to be on the cusp of a new literature that explores the phenomena and attempts to demonstrate the linkage between indicators and illustrate their impact with possible potential explanations.

4.4   Objectives of the Study

1. To study the conceptual framework of net interest margin (NIM) and its relevance with the economic growth.
2. To examine the linkages and impact of different macro- and bank-specific indicators on the earning capacity of public sector banks (PSBs) of Indian banking industry.

4.5   Research Methodology

4.5.1   Data Sources

The study is analytical and empirical in nature, which intends to demonstrate the linkages and impact of different macro- and bank-specific indicators on the earning capacity of PSBs of Indian banking industry. The present work is committed to improve the insight of regulators, supervisors, and investors and conclude with potential explanations. The econometric model developed for the study incorporates net interest margin (NIM) as a regress, and the predictors include lending interest rate (LIR), gross domestic product (GDP), inflation (INF), capital adequacy ratio (CAR), and non-performing loans (NPLs).

The data for lending interest rate, GDP, and inflation has been collected from the World Bank database, where as balance sheets, income statements, and their notes have been studied to get the data regarding CAR and NPL. The financial data has also been collected from the annual reports of the selected PSBs and publications of Reserve Bank of India (RBI) like Annual Report on Currency and Finance, RBI Bulletin, Financial Stability Report, and some information has also been brought from the relevant Web sites. The Government of India publications like Economic Survey 2014–15 and Union Budget 2014–15 have also been referred to gain better insight of economic indicators used in the study.

4.5.2   Nature of Data

In order to meet the research objectives, the study makes use of balanced panel data as it fits better than the single time series or cross sectional alone as advocated by

Brooks (2014). Further, the use of panel data tackles more complex problems than would be possible with pure time series or cross-sectional data. Brooks (2014) revealed that the combination of time series with cross sections can enhance the quality and quantity of data in ways that would be impossible using only one of these two dimensions.

4.5.3 Sample and Period of the Study

The data collected is a balanced pool of fifteen public sector banks in India, selected on the basis of market capitalization (National Stock Exchange). The study period is taken as fifteen financial years starting from April 01, 2001 to March 31, 2015. The scope of the study is limited only to the selected public sector banks excluding private sector banks and foreign banks operating in India.

4.5.4 Analytical Model

Keeping the cognizance of different perspectives of the study and past empirical literature, the study proposes the following classical multiple regression model:

$$nim_{it} = \beta_0 + \beta_1 lir_{it} + \beta_2 gdp_{it} + \beta_3 inf_{it} + \beta_4 crar_{it} + \beta_5 npl_{it} + e_{it}$$

where

β_0	The intercept of equation,
$\beta_1, \beta_2, \beta_3, \beta_4,$ **and** β_5	Coefficients for independent variables,
it	i represents the bank and t represents the year,
nim	Net interest margin,
lir	Lending interest rate,
gdp	Gross domestic product growth rate,
inf	Inflation,
crar	Capital adequacy ratio,
npl	Non-performing loans, and
e_{it}	Error or stochastic term.

For computation of variables, see Table 4.1.

4.5.5 Data Analyses Techniques

For the purpose of carrying out empirical analyses, this study utilizes econometric techniques, i.e., *Pooled OLS, Fixed Effects and Random Effects estimation models*

Table 4.1 Computation of variables

Variable	Measurement	Literature
Net interest margin (nim)	(Interest earned minus interest extended) divided by (total earning assets)	Ho and Saunders (1981), Saunders and Schumacher (2000), Claeys and Vander (2008), Horváth (2009), Schwaiger and Liebeg (2008)
Lending interest rate (lir)	Imported from World Bank	Maudos and De Guevara (2004), Brock and Suarez (2000), Saunders and Schumacher (2000)
Gross Domestic Product (gdp)	Annual percentage growth rate of GDP at market prices based on constant local currency	Alexiou and Sofoklis (2009), Demirgüç-Kunt and Huizinga (1999, Fadzlan and Kahazanah (2009), Claeys and Vander (2008), Horváth (2009)
Inflation (inf)	Inflation as measured by the annual growth rate of the GDP implicit deflator shows the rate of price change in the economy as a whole. The GDP deflator is the ratio of GDP in current local currency to GDP in constant local currency	Guru et al. (2002), Jiang et al. (2003), Perry (1992), Hoggarth et al. (2001), Tan and Floros (2012)
Capital adequacy ratio (crar)	(Tier I capital plus Tier II capital) divided by risk-weighted assets	Bourke (1989), Berger (1995), Angbazo (1997), Olaleken and Adeyinka (2013), Haslem (1969)
Non-performing loans (npl)	Ratio of the problem loans to total loans	Maudos and De Guevara (2004), Angbazo (1997)

Source Compiled by authors from different sources

related to balanced panel data. Balanced panel data is preferred over unbalanced panels, because it allows an observation of the same unit in every time period and reduces the noise introduced by unit heterogeneity (Brooks 2014). The classical linear multiple regression model establishes the relationship between the variables under study. For the effective analysis of the data, MS Excel and econometric package *STATA 13* have been used.

Before carrying out the analysis, linear multiple regression model was carried out under the *Pooled OLS, Fixed Effects and Random Effects Estimation Models*. To check as to which model is fit and appropriate between fixed and random effects model, *Hausman Test* identified random effects model as an appropriate model. To differentiate between Pooled OLS and the random effects model, *Breusch–Pagan Lagrange Multiplier Test* was applied which declared random effects model to be the appropriate model. Further, it is mandatory to fulfill the assumptions of the model, viz. multicollinearity, heteroscedasticity, and autocorrelation in particular. The model calculates variance inflation factors and tolerance values and also plots the correlation matrix to detect the chances of multicollinearity. Secondly, the model is tested for heteroscedasticity by *Breusch–Pagan* or *Cook–Weisberg Test for Heteroscedasticity.* Thirdly, autocorrelation in panel data was checked through *Wooldridge Test of Autocorrelation.*

4.6 Analysis

The model summary of the data is highlighted in Table 4.2. From Table 4.2, the NIM of the sample banks has ranged from 0.94 to 8.92% with a standard deviation and average of 1.03 and 3.47%, respectively. From the macroeconomic environment, lending interest rates have varied from 8.3 to 14% with a standard deviation and average of 1.45 and 11.44%, respectively, whereas Indian economic growth has ranged from 3.8 to 10.3% with a standard deviation and average of 2.20 and 7.02%, respectively. Similarly, from the bank-level indicators, CRAR was recorded minimum at 1% in the year 2001 to maximum of 20.11% in the year 2004 with a standard deviation and average of 1.96 and 12.28%, respectively. With regard to non-performing loans, minimum value of 2.2% and maximum value of 12.8% were registered along with a standard deviation and mean value of 3.51 and 5.56%, respectively.

Table 4.3 highlights the correlation between the variables used in the study. As is evident from Table 4.3, pair-wise correlation coefficients indicate statistically significant positive correlation of net interest margin with lending interest rates and capital to risk-weighted assets ratio, whereas a negative correlation of NIM with GDP, INF, and NPL is observed at 5% level of significance. The correlation matrix

Table 4.2 Descriptive statistics

Variable	Obs.	Mean	Standard Deviation	Min	Max
Nim	225	3.47	1.03	0.94	8.92
Lir	225	11.44	1.45	8.3	14
Gdp	225	7.02	2.20	3.8	10.3
Inf	225	5.626667	1.805671	3.2	9
Crar	225	12.28	1.96	1	20.11
Npl	225	5.56	3.51	2.2	12.8

Source Results Obtained from using STATA Software

Table 4.3 Correlation matrix

Variables	Nim	Lir	Gdp	Inf	Crar	Npl
Nim	1.00					
Lir	0.22	1.00				
Gdp	−0.05	−0.52	1.0			
Inf	−0.31	−0.38	0.22	1.00		
Crar	0.06	−0.31	0.30	0.32	1.00	
Npl	−0.46	0.47	−0.53	−0.76	−0.38	1.00

Source Results Obtained from using STATA Software

Table 4.4 Collinearity statistics

Variable	VIF[b]	Tolerance[a] (1/VIF)
Lir	1.55	0.64
Gdp	1.92	0.52
Inf	2.87	0.35
Crar	1.21	0.83
Npl	3.70	0.27
Mean VIF	2.25	

[a]Most commonly tolerance values of 0.10 or less are cited as problematic
[b]*VIF* Variance inflation factor (VIF, stands for variance inflation factor, is the reciprocal of tolerance. It indicates the degree to which the standard errors are inflated due the level of collinearity. Most commonly VIF values more than 5 are cited as problematic.)
Source Results obtained using STATA software

also reveals correlation between independent variables, where NPL is negatively associated with GDP, INF, and CRAR which are positively associated with LIR. In the same way, CRAR is positively correlated with GDP and INF, while it is negatively associated with LIR. Similarly, INF is found to be negatively associated with LIR and positively correlated with GDP. As far as GDP is concerned, it is negatively associated with LIR. From Table 4.3, high correlation of 76% between INF and NPL signals the chances of multicollinearity. Therefore, to detect the chances of multicollinearity in order to avoid biased regression coefficients, due care has been taken in the regression analysis by evaluating the tolerance levels and variance inflation factor (VIF) values of the independent variables used in this study.

Table 4.4 presents the calculated VIF and tolerance values to detect any chances of multicollinearity among the independent variables. The VIF values, as is elicit from the observations, range from 1.21 to 3.70, which clearly verifies that none of the observation surpasses the threshold limit of 5 whether individually or collectively (Mean VIF = 2.25). This conclusion is also supported by the values of tolerance, as none of the observation is nearing zero. Consequently, it is concluded that there is no problem of multicollinearity between the independent variables considered under the study (Gujarati and Porter 2009; Marquardt 1970).

Another problem commonly encountered in the cross-sectional data is heteroscedasticity, i.e., unequal variance in the error term. The problem is diagnosed by using *Breusch–Pagan* or *Cook–Weisberg Test for Heteroscedasticity*[1] which verified the absence of homoscedasticity (equal variance) as the *p-value*

---

[1]*Breusch–Pagan/Cook–Weisberg test for heteroscedasticity (STATA output)*:
H_0: Constant Variance; Variables: fitted values of NIM
chi2(1) = 189.64
Prob > chi2 = 0.000.

(0.000) for the test was found significant, thus rejecting the null hypothesis of constant variance. To overcome this problem, *Robust Standard Errors* have been used for the reason; OLS assumes that errors are both independent and identically distributed. Therefore, robust standard errors relax either or both of those assumptions. Hence, when a symptom of heteroscedasticity is diagnosed, robust standard errors tend to be the best treatment and therefore, more trustworthy (Williams 2015).

After the assumptions of multicollinearity and heteroscedasticity, serial correlation was also examined for the model, which is a common problem found in time series data. For the panel data, serial correlation also called as autocorrelation was diagnosed by applying *Wooldridge Test of Autocorrelation.*[2] The results plotted by the test verified the presence of autocorrelation as the significant *p-value (0.004)* for the test rejected the null hypothesis of no serial correlation. First difference operator has been used, which helped in overcoming the problem. But the result obtained after first differencing the variables is assumed to produce less reliable results as the procedure ignores the first observation, which may not matter for large samples but may prove detrimental in case of small samples. In order to cure this problem, another procedure called as *Prais–Winsten Transformation* (Gujarati 2014) has been used to take into account the first observation as well. Therefore, the results portrayed by the Prais–Winsten estimation indicated a transformed *d-statistic*[3] of 1.76, which lies between the upper bound (dU) of 1.82 and lower bound (dL) of 1.70, thus signaling absence of serial correlation in the model (Table 4.5).

4.6.1 Interpretation and Findings

Table 4.6 shows the result of multiple regression analysis carried on a sample of 15 PSBs for a period from April 2001 to March 2015 comprising of five predictors regressed against NIM (dependent variable). The value of R^2 in the model is estimated at 0.36 which reveals that predictors collectively are responsible for bringing a change of 36% in NIM. The value of *0.000* for *prob > chi2* indicates that the model is nicely fitted as none of the regression coefficients for explanatory variables equals zero. The regression coefficients for each regressor along with their p-values and standard errors are also highlighted in Table 4.6.

It can be estimated that there exists a positive and significant correlation between NIM and LIR as the values of regression coefficient and p-value are 0.1347 and 0.001, respectively. It means that there will be positive change of 13.47% in NIM of

---

[2]*Wooldridge test for autocorrelation in panel data (STATA output)*:

   H_0: no first-order autocorrelation

   $F (1, 14) = 1.448$

   Prob $> F = 0.004$.

[3]The values of dL and dU are taken from Durbin–Watson table for 225 observations with 5 regressors at 5% level of significance (where d-statistic stands for Durbin–Watson Statistic).

Table 4.5   Prais–Winsten transformation

```
Number of gaps in sample:   14     (gap count includes panel changes)
(note:  computations for rho restarted at each gap)

Interation 0: rho   =    0.0000
Interation 1: rho   =    0.5119
Interation 2: rho   =    0.5719
Interation 3: rho   =    0.5773
Interation 4: rho   =    0.5778
Interation 5: rho   =    0.5778
Interation 6: rho   =    0.5778
Interation 7: rho   =    0.5778

Prais - Winsten AR(1) regression - iterated estimates
```

Source	SS	df	MS			
				Number of obs	=	225
				F(5, 219)	=	35.14
Model	76.5840135	5	15.3168027	Prob > F	=	0.0000
Residual	95.4555487	219	.435870085	R - squared	=	0.4452
				Adj R - squared	=	0.4325
Total	172.039562	224	.76803376	Root MSE	=	.6602

nim	Coef.	Std. Err.	t	p>|t|	[95% Conf. Interval]	
lir	.0785116	.0481018	1.63	0.104	-.0162901	.1733133
gdp	.0876792	.027873	3.15	0.002	.0327456	.1426128
inf	.1351998	.0359913	3.76	0.000	.0642662	.2061333
crar	.0997439	.034306	2.91	0.004	.0321317	.1673562
npl	-.2107187	.0286048	-7.37	0.000	.1543429	.2670946
_cons	-1.254798	.9668029	-1.30	0.196	-3.160226	.6506311
rho	.5778216					

```
Durbin - Watson statistic  (original)       0.765563
Durbin - Watson statistic  (transformed)    1.728590
```

Source: Results obtained using STATA Software

Table 4.6   Regression analysis

```
Random - effects GLS regression          Number of obs      =   225
Group variable:     id                   Number of groups   =   15

R-sq:  Within    =    0.4539             Obs per group:Min  =   15
       Between   =    0.0153                         Avg   =   15.0
       Overall   =    0.3648                         Max   =   15
                                         Wald chi2(5)  =   170.86
corr(u_i, X)    =    0 (assumed)          Prob > chi2   =   0.0000
```

nim	Coef.	Std. Err.	z	p>|z|	[95% Conf. Interval]	
lir	.1347175	.0412731	3.26	0.001	.0538237	.2156112
gdp	.1806801	.0301886	5.99	0.000	.1215116	.2398486
inf	.1713598	.0448926	3.82	0.000	.0833719	.2593477
crar	.1435962	.02995	4.79	0.000	.0848952	.2022972
npl	-.2690131	.0262464	10.25	0.000	.2175711	.3204551
-cons	-3.56515	.8741003	-4.08	0.000	-5.278355	-1.851945
sigma_u	.44678155					
sigma_e	.71702587					
rho	.27967299	(fraction of variance due to u_i)				

Predictors: lir - Lending Interest Rate, **gdp** – Gross Domestic Product Growth Rate, **inf** - Inflation,

crar - Capital to Risk Weighted Assets Ratio and **npl** – Non Performing Loans.

Dependent Variable: nim - Net Interest Margin.

Source: Results obtained using STATA Software.

PSBs as a result of a unit change in LIR. In other words, bank's policy of increasing the lending interest rates will prove improving the NIM. Among the macroeconomic indicators, there exists a positive and statistically direct relationship between NIM and GDP as the values for regression coefficient and p-value are 0.1807 and 0.000, respectively. Similarly, direct and significant relationship between INF and NIM can be estimated from the regression coefficient and p-value, which stand at 0.1713 and 0.000%, respectively. These values confirm that a unit change in inflation can bring a positive variation of 17.13% in NIM. From the bank-level indicators, a positive correlation of CRAR and NIM can be estimated as the regression coefficient stands at 0.1436% (approx.). Further, this relationship is statistically significant as the p-value stands at 0.000. All the PSBs have maintained CRAR on an average well above the threshold value of 9% as laid down by Basel II. Moreover, a negative and statistically significant association between NPL and NIM can be estimated as is elicit from the regression coefficient of −0.2690% and p-value of 0.000. This negative correlation signals that 26.90% downside change in NIM can be experienced, if a unit change in NPL is observed.

Though all the relationships are economically significant, it can also be observed from Table 4.6 that except NPL, all other predictors bear a positive and significant relationship with NIM. For brevity of exposition, these statistically and economically significant correlations are consistent with the theoretical literature and variably related to the empirical evidences.

All macroeconomic indicators included in the study proved to be statistically and directly linked to NIM, meaning that macroeconomic environment in which PSBs operate, significantly influences their performance. The positive connection between NIM and GDP implies that in the periods of higher growth, NIM can increase due to proliferation in the credit activity and better loan quality. The PSBs can improve their efficiency by augmenting their NIM, when they operate in the favorable macroeconomic ambience as verified by the empirical findings related to INF and GDP. There are various empirical studies like Tan and Floros (2012), Fadzlan and Kahazanah (2009), Jiang et al. (2003), Guru et al. (2002), and Demirgüç-Kunt and Huizinga (1999) which support this inference. Further, the positive economic growth can be understood as the productive functioning of all the sectors in an economy, where banking is not an exception. With respect to inflation, PSBs are seen benefitting from the increase in inflation in terms of NIM. The possible reason for the relationship can be conceived from the impact of inflationary pressures on the economy. To grapple such pressures, banks normally employ the policy of increasing lending rates in order to limit the circulation of money into the economy and bring down the inflation to the normal level. The RBI executes the strategy of contractionary monetary policy by either allowing banks to expand the money supply more slowly than usual or strictly directing them to shrink the supply of money for the sake of price stability and thus financial stability. During such periods, any advances made by the PSBs can boost their efficiency by experiencing higher NIM through increased lending rates. Laconically, it can be thought of that increased lending rates especially during the periods of higher inflation can prove beneficial for the PSBs. Thus, the results are in line with Perry (1992).

Within the banking environment, a significant negative correlation observed between NPL and NIM signals about stressed financial statements of PSBs. The underlying possible explanation can be attributed to lending and credit extension by the PSBs to the private sector, which resulted in the deteriorated picture of their balance sheet and erosion of their capital. The problem of NPL has become a stumbling block in the path of income generation and in beautifying the capital base of PSBs. The languished picture of financial statements or balance sheet syndrome has further led to inadequate loss provisions resulting into the adverse impact on depositor's and investor's confidence. This way PSBs are apprehended to be prone to the financial risks, stagger and direct themselves to get hobbled by the reputational risk. This catchall of circumstances in the PSBs emphasizes to institute better policies of lending and credit to rejuvenate their actual potential and thus invigorate the economy. Therefore, PSBs can combat the problem of NPL by earmarking reserves of reasonable magnitude and periodically keep monitoring their long-term debtors for income smoothing. The adverse consequences of rising NPLs may otherwise infuse credit risk, inviting liquidity crisis and thus, eroding the capital buffers as well.

With regard to CRAR, it has proved to be a cushion for the survival of PSBs as usual, as the PSBs are very prudent in maintaining capital ratio more than profit making. This risk aversion mechanism in terms of CRAR may signal the conservative behavior adopted by PSBs. This may also be estimated that due to the rising toll of NPAs, regulators start weaving safety nets through increased buffers of capital with the opportunity cost of missing the investment avenues to elevate their efficiency through financial intermediation. These inferences are also supported by empirical evidences like Bourke (1989), Berger (1995), Angbazo (1997), Haslem (1969), and Olalekan and Adeyinka (2013).

4.7 Limitations of the Study

The study is confined to a time period of fifteen years from 2000–01 to 2014–15 and includes only public sector banks, comprising of fifteen representative banks on the basis of market capitalization (NSE), which as a whole constitutes 56% of Indian public sector banks. Furthermore, only a few indicators from macro- and bank environment have been studied on the grounds of their criticality, relevance, and understandability.

4.8 Conclusion

In this study, we analyzed the determinants of NIM from both macroeconomic and bank environments for a sample of 15 PSBs during the period starting from April 2001 to March 2015. The study used classical multiple linear regression with fixed effects and random effects model. The regression analysis educed with a host of conclusive remarks and policy implications.

As far as macroeconomic performance is taken into account, it significantly influences the financial intermediation of PSBs. The favorable economic ambience can prove as a main driver for encouraging net interest margin of PSBs, which can be traced from the empirical relevance of inflation and GDP with the NIM. The growing economy implies smooth functioning of private sector, which can be seen as an indication of the recovery of bad debts pertaining to the PSBS. Inflation as a concomitant element of the economic growth can be put under the thumb by the timely ensured monetary policies. In light of favorable macroeconomic circumstances, PSBs can efficiently intermediate but as the inflation paces up, contractionary monetary policy can prove benign for the PSBs. As the policy increases, the short-term lending rates, the borrowing capacity and attractiveness of loans diminish. During the time, any advances made by the PSBs give impetus to their net interest margins. High capital to risk-weighted assets ratio (CRAR) has implied the prudence, risk aversion behavior, and also foregoing of profitable investment opportunities on the part of PSBs. The PSBs need to revamp their policy with respect to CRAR as unreasonable or excessive quantum of capital buffers will only make them incur opportunity cost of relinquishing the lucrative investment avenues.

As far as NPLs are concerned, it has proved to be a chronic problem especially with PSBs in India. The policy implications like cutting the loan loss provisions to the level of reasonability, timely and consistent implementation of guidelines from the central bank, cautious credit extension, persistent monitoring of long-term debts, and accessing Reserve Bank Of India's database on non-performing loans (CRILC—Credit Repository Information On Large Credits) will be instrumental in bringing the NPLs down to a large possible extent.

It is suggested for the public sector banks to merge for better consolidation, allocation of funds, and investment prospects. It is also suggested to install better risk management practices and let them trickle down to the branch level for better monitoring of interest rate structure, capital adequacy norms, and NPL management along with other operations.

4.9 Directions for Future Research

As the current study is limited to a few PSBs, researchers may enquire into the same phenomena for all public sector banks for substantial results. Further areas for research may also include impact of interest rate risk and credit risk on NIM of banks; studying the behavior of NIM during recessionary and booming periods in an economy alongside inflationary pressures on the earnings of banks can also be the debatable topics for future research. Researchers may also deliberate on issues like non-performing loans and liquidity crisis particularly in public sector banks of India.

References

Abreu, M., & Mendes, V. (2001). Commercial bank interest margins and profitability: Evidence for some EU countries. In *Pan-European Conference Jointly Organized by the IEFS-UK & University of Macedonia Economic & Social Sciences, Thessaloniki, Greece, May* (pp. 17–20).

Alexiou, C., & Sofoklis, V. (2009). Determinants of bank profitability: Evidence from the Greek banking sector. *Economic Annals, 54*(182), 93–118.

Altunbarş, Y., Molyneux, P., & Thornton, J. (1997). Big-Bank mergers in Europe: An analysis of the cost implications. *Economica, 64*(254), 317–329.

Angbazo, L. (1997). Commercial bank net interest margins, default risk, interest-rate risk, and off-balance sheet banking. *Journal of Banking & Finance, 21*(1), 55–87.

Beck, T., & Hesse, H. (2009). Why are interest spreads so high in Uganda? *Journal of Development Economics, 88*(2), 192–204.

Berger, A. N. (1995). The relationship between capital and earnings in banking. *Journal of Money, Credit and Banking, 27*(2), 432–456.

Bourke, P. (1989). Concentration and other determinants of bank profitability in Europe, North America and Australia. *Journal of Banking & Finance, 13*(1), 65–79.

Brock, P. L., & Suarez, L. R. (2000). Understanding the behavior of bank spreads in Latin America. *Journal of Development Economics, 63*(1), 113–134.

Brooks, C. (2014). *Introductory econometrics for finance*. Cambridge University Press.

Busch, R., & Memmel, C. (2017). Banks' net interest margin and the level of interest rates. *Credit and Capital Markets-Kredit und Kapital, 50*(3), 363–392.

Claeys, S., & Vander, V. R. (2008). Determinants of bank interest margins in Central and Eastern Europe: A comparison with the West. *Economic Systems, 32*(2), 197–216.

Demirgüç-Kunt, A., & Huizinga, H. (1999). Determinants of commercial bank interest margins and profitability: Some international evidence. *The World Bank Economic Review, 13*(2), 379–408.

Dhal, S., Kumar, P., & Ansari, J. (2011). Financial stability, economic growth, inflation and monetary policy linkages in India: An Empirical Reflection. *Reserve Bank of India Occasional Papers, 32*(3), 1–35.

Economic Survey (2015). Ministry of Finance, Government of India. Retrieved at http://indiabudget.nic.in/es2014–15/echapter-vol1.pdf.

Fadzlan, S., & Kahazanah, N. B. (2009). Determinants of bank profitability in a developing economy: Empirical evidence from the China banking sector. *Journal of Asia-Pacific Business, 10*(4), 201–307.

Financial Stability Report. (2015). Reserve Bank of India. Retrieved at https://rbidocs.rbi.org.in/rdocs/PublicationReport/Pdfs/0FSR6F7E7BC6C14F42E99568A80D.

García-Herrero, A., Gavilá, S., & Santabárbara, D. (2009). What explains the low profitability of Chinese banks? *Journal of Banking & Finance, 33*(11), 2080–2092.

Georgievska, L., Kabashi, R., Manova-Trajkovska, N., Mitreska, A., & Vaskov, M. (2011). Determinants of lending interest rates and interest rates spreads. In *InBank of Greece, Special Conference Paper* (Vol. 9).

Goyal, K. A., & Joshi, V. (2012). Indian banking industry: Challenges and opportunities. *International Journal of Business Research and Management, 3*(1), 18–28.

Gujarati, D. (2014). *Econometrics by example*. Palgrave Macmillan.

Gujarati, D. N., & Porter, D. C. (2009). *Basic Econometrics* (5th ed.). Boston: McGraw-Hill.

Gup, B. E., & Kolari, J. W. (2006). *Commercial Banking: The Management of Risk*. Wiley Incorporated.

Guru, B. K., Staunton, J., & Balashanmugam, B. (2002). Determinants of commercial bank profitability in Malaysia. *Journal of Money, Credit, and Banking, 17*(1), 69–82.

Haque, S. M., & Wani, A. A. (2015). Relevance of financial risk with financial performance: An insight of Indian banking sector. *Pacific Business Review International, 8*(5), 54–64.

Haslem, J. A. (1969). A statistical estimation of commercial bank profitability. *The Journal of Business, 42*(1), 22–35.

Hoggarth, G., Milne, A., & Wood, G. E. (2001). Alternative routes to banking stability: A comparison of UK and German banking systems. In S. Frowen & F. Machugh (Eds.), *Financial competition, risk and accountability* (pp. 11–32). UK: Palgrave Macmillan.

Horváth, R. (2009). *Interest margins Determinants of Czech banks* (No. 2009/11). Charles University Prague, Faculty of Social Sciences, Institute of Economic Studies.

Ho, T. S., & Saunders, A. (1981). The determinants of bank interest margins: Theory and empirical evidence. *Journal of Financial and Quantitative Analysis, 16*(4), 581–600.

Jiang, G., Tang, N., Law, E., & Sze, A. (2003). Determinants of bank profitability in Hong Kong. *Hong Kong Monetary Authority Research Memorandum, 6,* 2015.

Kanwal, S., & Nadeem, M. (2013). The impact of macroeconomic variables on the profitability of listed commercial banks in Pakistan. *European Journal of Business and Social Sciences, 2*(9), 186–201.

Kasman, A., Tunc, G., Vardar, G., & Okan, B. (2010). Consolidation and commercial bank net interest margins: Evidence from the old and new European Union members and candidate countries. *Economic Modelling, 27*(3), 648–655.

Khan, W. A., & Sattar, A. (2014). Impact of interest rate changes on the profitability of four major commercial banks in Pakistan. *International Journal of Accounting and Financial Reporting, 4* (1), 142.

King, R. G., & Levine, R. (1993). Finance and growth: Schumpeter might be right. *The Quarterly Journal of Economics, 108*(3), 717–737.

Marquaridt, D. W. (1970). Generalized inverses, ridge regression, biased linear estimation, and nonlinear estimation. *Technometrics, 12*(3), 591–612.

Maudos, J., & De Guevara, J. F. (2004). Factors explaining the interest margin in the banking sectors of the European Union. *Journal of Banking & Finance, 28*(9), 2259–2281.

Mishra, R. N., Majumdar, S. & Bhandia, D. (2013). *Banking stability—A precursor to financial stability.* Retrieved at http://rbidocs.rbi.org.in/rdocs/Publications/PDFs/1WPS18012013.PDF.

Olalekan, A., & Adeyinka, S. (2013). Capital adequacy and banks' profitability: An empirical evidence from Nigeria. *American International Journal of Contemporary Research, 3*(10), 87–93.

Perry, P. (1992). Do banks gain or lose from inflation? *Journal of Retail Banking, 14*(2), 25–31.

Poghosyan, T. (2013). Financial intermediation costs in low income countries: The role of regulatory, institutional, and macroeconomic factors. *Economic Systems, 37*(1), 92–110.

Rousseau, P. L., & Wachtel, P. (1998). Financial intermediation and economic performance: Historical evidence from five industrialized countries. *Journal of Money, Credit and Banking*, 657–678.

Saunders, A., & Schumacher, L. (2000). The determinants of bank interest rate margins: An international study. *Journal of International Money and Finance, 19*(6), 813–832.

Schwaiger, M. S., & Liebeg, D. (2008). Determinants of bank interest margins in Central and Eastern Europe. *Financial Stability Report, 14*(1), 68–87.

Tan, Y., & Floros, C. (2012). Bank profitability and inflation: The case of China. *Journal of Economic Studies, 39*(6), 675–696.

Williams, R. (2015). *Heteroskedasticity.* Available at https://www3.nd.edu/~rwilliam/stats2/l25.pdf.

Chapter 5
A Trend Analysis of Reforms in the Indian Bond Market

Rahul Rangotra

Abstract The paper analyses the impact of various reforms undertaken by the government of india to improve liquidity, transparency, and security in the Indian bond market. It considers reforms initiated by government of india since 1992 that include introduction of system of primary dealers, establishment of Clearing Corporation of India Limited as a clearinghouse, introduction of screen-based trading in government securities through negotiated dealing system-order matching (NDS-OM), trading of bonds through stock exchanges, introduction of delivery versus payment system, etc. Time series graphs are used for analysis by collecting secondary data from Reserve Bank of India, Securities and Exchange Board of India, Clearing Corporation of India Limited, and National Stock Exchange. Indian government securities market has changed significantly in the last two decades. The impact of reforms on the Indian bond market is examined by analyzing the combined gross borrowing of center and state government through government securities (increased by around 8900% from 1991–92 to 2016–17), secondary market transactions in government securities (increased by around 430,000% from September 1994 to September 2017), net corporate debt outstanding (increased by around 225% from June 2010 to September 2017), total trade in corporate bond market (increased by around 1450% from 2007–08 to 2016–17), and other variables related to the liquidity and size of Indian bond market. The impact of reforms is found to be positive for all the dimensions but have significant impact only on the size and liquidity of the Indian bond market. The study concludes with strategic implications.

Keywords Corporate bond market · Government securities market India

R. Rangotra (✉)
Department of Management Studies, Central University of Kashmir,
Srinagar, Jammu and Kashmir, India
e-mail: rahulrangotra@gmail.com

5.1   Introduction

Before eighteen century, the Indian princely states used to meet the borrowing requirements from indigenous bankers. Raising debt from the public was first introduced by the East India Company to finance Anglo-French wars (RBI, n.da). Raising the public debt was also one of the main reasons for setting central bank in India. In 1867, first time the public debt was raised to finance railways construction and public works like irrigation canals in India. British government also raised public debt to finance the cost of war. In India, public debt was managed by Comptroller and Auditor General of India till 1913 and by controller of currency till 1935. After the Reserve Bank of India commenced its operations, the public debt is managed by the public debt office of RBI. After independence of India the public debt was used to finance the five years plans. Between 1985 and 1991 with the recommendation of the Chakravarty Committee Report, the attempt was made to align the interest rate on government securities with market interest rates.

Indian bond market is an emerging bond market. However, it is comparatively small and less liquid than developed bond markets and is ranked fourth in Asia with 70% share of government bonds, (Sabnavis and Mehta 2014). Relative to its Asian peers, a wide range of issuers, such as government, public sector undertakings, banks and corporates and investors participate in Indian bond market and due to underdeveloped corporate bond market in emerging markets, corporations generally depend on the banks to raise capital. This consequently results in more credit risk to be borne by the banks. To overcome the problems in the Indian bond market, the government of india has introduced various reforms since 1992. Before 1992, the market of government securities was dominated by the banks due to high statutory liquid ratio (SLR) and cash reserve ratio (CRR). The interest rates were kept very low to provide low-cost finance to the government. Liquidity was lacking, secondary market was not transparent, and there was lack of smooth yield curve which can be used as a benchmark for corporate bond market.

5.2   Review of Literature

The corporate debt market is important for economy and its stakeholders like government, corporate, investors, and financial institutions (The International Capital Markets Association 2013). As revealed by many studies that relationship between the size of corporate bond market, equity market, bank loan, and government bond market is vital for the development of the corporate bond market and act as benchmark for the corporate bond market. Endo (2000) argued that corporate bond market can substitute the long-term bank loans and reduces the risk in the banking system. Goodfriend (2005) argued that the interest cost is less for the firms which raise funds from the corporate bond market. Endo (2000) observed that in developing countries the corporate bond market can reduce the risk of banking system and regulatory, and institutional support is required for corporate bond market. Rakshit (2000), Herring et al. (2000), Corsetti et al. (1998) argued that had there been bond market particularly corporate bond market the Asian financial crisis of 1997 would have been avoided. Wells and Schou-Zibell (2008) argued that for the development of corporate bond market regulatory framework is important and disorganized and inconsistent regulatory framework hinders the development of the corporate bond market. Endo (2000) argued that the reforms in the corporate bond market were started in the developed bond markets like USA, Germany, Japan after the reforms in the equity and the government bond market (Endo 2000). Wells and Schou-Zibell (2008) stated that the securitization in India is yet to be taken off. Luengnaruemitchai and Ong (2005) argue that the crowding out increases the cost of borrowing for the corporate. But in India it is otherwise, i.e., government bond market helped in the development of the corporate bond market (Raghavan and Sarwano 2012). The banking system is large in the economy where the corporate bond market is absent (Harkansson 1999). So, analyzing the impact of reforms in the Indian government securities and corporate bond market is important for at least five reasons. First, government securities markets help to fund the budget deficit of the government in a non-inflationary way. Second, it improves the effectiveness of monetary policy of the central bank. Third, sovereign yield provides the benchmark for valuation of other securities. Fabella and Madhur (2003) suggested that active sovereign bond market need to be established which serves as a base for the development of corporate bond market. Particularly, sovereign debt market of a country should be efficient, liquid, transparent, and risk-free to act as benchmark in valuation of other securities in the country. Fourth, research on the factors affecting bond market is important to help the policymakers to improve the stability and efficiency of the financial system. Fifth, domestic bond market needs to be developed as it helps in less dependency on foreign currency debt which causes problems during fluctuations in exchange rate. It also helps in efficient operations of monetary policy. Also, development of corporate bond market is necessary to reduce the dependence on the banking system and diversify the credit risk in the economy. Taking into consideration the significance of the bond markets the present study is conducted to analyze the impact of reforms undertaken by the government of india since 1992 on the Indian bond market.

The paper is divided into sections. The second following section of the paper discusses the reforms in the Indian government securities market since 1992, third section highlights the research methodology used, fourth section shows the impact of reforms in Indian government securities and corporate bond market with the help of graphs, and the last section gives the concluding remarks.

5.3   Reforms in the Indian Bond Market

The Indian debt market can be divided into three main segments: government securities market, public sector undertaking bonds, and corporate debt market. In secondary market, the government securities can be traded through negotiated dealing system-order matching (NDS-OM) and through telephone/over the counter. Clearing Corporation of India Limited operates NDS-OM. Stock exchanges like National Stock Exchange, Bombay Stock Exchange are allowed to facilitate trading in government securities, public sector undertaking bonds and corporate bonds. These exchanges have separate debt market segments called as wholesale debt market and corporate bond market. National Securities Depository Limited (NSDL) and Central Depository Services Limited (CDSL) act as depositaries in case of government securities traded through stock exchanges. In stock exchanges, these securities are regulated by Securities and Exchange Board of India (SEBI). There is long list of reforms in the Indian bond market since 1992 (RBI 2007).

Reforms in the government securities and corporate bond market in India were started in 1992. Before 1992, government securities market was dominated by the banks due to various factors such as high statutory liquid ratio (SLR), low-interest rates, low liquidity, and transparency in the secondary bond market, and due to all these, there is lack of smooth yield curve which can be used as a benchmark. Mohan (2004a, b) in his paper remarked even after reforms that due to high cash reserve ratio (CRR) and SLR. RBI has very little room for monetary maneuvering. The government securities market has undergone the reforms such as adoption of screen-based trading, holding of government securities in dematerialized form, incorporation of Clearing Corporation of India Limited, issuing of new types of government securities like inflation-indexed bonds, securities with embedded options, floating rate government securities, deregulation of interest rates, improvement in transparency and liquidity, better legal environment. Further, the network of primary dealers was introduced and wholesale debt market segment was introduced in National Stock Exchange and Bombay Stock Exchange. Also, retail and foreign investors are entering into the market beside the Indian institutional investors. (Sabnavis and Mehta 2014) remarked that in India among the different institutional investors, banks are the major investors of government bonds in India. Domestic financing is dominated by the bank credit, and the share of corporate bonds is only 18.4% while foreign holding in local government bonds is only 1.4%.

Reforms in the bond market were essential due to many reasons, some of the reasons were (a) to finance the increasing budget deficit of the government at reasonable cost, (b) to increase the investor base, (c) to increase the overall efficiency of the Indian capital market, (d) to improve the liquidity and transparency in the secondary market of the government securities, (e) to help in funding the infrastructure projects, (f) to better implementation of monetary policy, (g) to reduce the burden of lending and share the credit risk with the banks, (h) to help the development of corporate bond market, as government securities' market act as a benchmark for corporate bond market, (i) to help the corporates to raise debt from the market efficiently.

5.4 Methodology

The objective of the paper is to analyze the impact of reforms undertaken by the government of india since 1992 in the Indian bond market. The secondary data was collected from Reserve Bank of India (RBI, n.db), Securities and Exchange Board of India, Clearing Corporation of India Limited, and National Stock Exchange. All the data was analyzed under seven heads, namely expansion of the government securities market (from 1980 to 2017), expansion of Indian corporate bond market (from 2009 to 2017), turnover of government securities market (from 1998 to 2017), treasury bills outstanding (from 2006 to 2017), ownership pattern of the government of india dated securities (from 2007 to 2017), maturity pattern of government securities (from 1976 to 2017), and Indian 10-year government bond yield (from 1994 to 2017). Period of the analyzing depends upon the availability of the data. Time series graphs and descriptive statistics are used for data analysis.

5.5 Impact of Reforms on Bond Market

5.5.1 Expansion of Governments Securities Market

Indian government finances its fiscal deficit through debt financing by issuing short-term, medium-term, and long-term securities. Table 5.1 shows the gross fiscal deficit (GFD) of central government and its financing through market borrowings. The gross fiscal deficit has increased from Rs. 82.99 billion in 1980–81 to Rs. 5465.32 billion in 2017–18, and the major portion of this fiscal deficit was financed by the market borrowing, i.e., by issuing of government securities. As it is clear from Table 5.1 and Fig. 5.1 that major portion of the GFD of the government of india is financed by issuing the government securities. The percentage of GFD financed from market borrowings has increased from 32% in the year 1980–81 to 63% in the year 2017–18. This is a significant change.

State governments also raise funds by issuing government bonds, called as state development loans (SDL), with the help RBI to finance their budget deficit. SDL add to the supply of government securities in the market, these securities also eligible for investment under statutory liquid ratio (SLR). Table 5.2 shows the states' gross fiscal deficit and its financing, and it is clear from Table 5.2 and Fig. 5.2 that the major portion of the state gross fiscal deficit is also financed by issuing government securities, i.e., market borrowings. The percentage of GFD

R. Rangotra

Table 5.1 Centre's gross fiscal deficit of central government and its financing (Rupees billion)

Year	1980–81	1985–86	1990–91	1995–96	2000–01	2005–06	2010–11	2015–16	2017–18
Gross fiscal deficit	82.99	218.58	446.32	602.43	1188.16	1464.35	3735.91	5327.91	5465.32
Financing of GFD—market borrowings	26.79	48.84	80.01	340.01	734.31	1062.41	3263.99	4149.31	3482.26
% of GFD financed from market borrowings	32.281	22.34,422	17.9266	56.43975	61.80228	72.55164	87.368	77.87876	63.71557

Source of data RBI's Database on Indian Economy

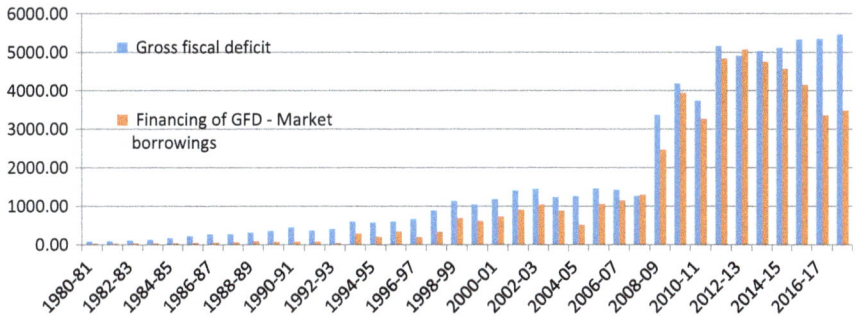

Source of Data: RBI's Database on Indian Economy

Fig. 5.1 Centre's gross fiscal deficit of central government and its financing

financed from market borrowings has increased from 5% in the year 1980–81 to 75% in the year 2017–18, beside this total financing of GFD through market borrowing has increased from 1.98 billion rupees to 3387.42 billion rupees, respectively (Fig. 5.3).

Gross fiscal deficit as percentage of GDP has decreased but the quantum of deficit has increased as the Indian economy is expanding. Since 1970, gross fiscal deficit of central government is fluctuating between 2.5 and 8% of GDP. In the current fiscal year, gross fiscal deficit is estimated as 3.24% of GDP. Table 5.3 and Fig. 5.4 show the combined fiscal deficit of state and central government. The combined fiscal deficit is very high even if the gross fiscal deficit of the central government is decreasing. The combined gross fiscal deficit in the current fiscal is 6.5% of GDP. All this show that both state and central government have to depend on the market borrowing for financing the fiscal deficit which increases the supply of the government securities in the Indian government securities market.

Table 5.3 and Fig. 5.5 show the combined market borrowings of central and state governments since 1980. The combined market borrowings of central and state governments are increasing and in the year 2016–2017 reached to 11,065.05 billion rupees gross and 7393.41 billion rupees net.

This shows that the supply of the Indian government securities is increasing very fast. As India is a developing country, deficit budget is the requirement for development of the economy and to finance the fiscal deficit government has to depend on the market borrowing. Because of this the supply of government securities has increased significantly.

5.5.2   Expansion of Indian Corporate Bond Market

Indian corporate bond market is small as compared to the government securities market. Corporate bond market has expanded many times after the reforms.

Table 5.2 States' gross fiscal deficit and its financing (Rupees billion)

Year	1980–81	1985–86	1990–91	1995–96	2000–01	2005–06	2010–11	2015–16	2016–17
Gross fiscal deficit	37.13	75.21	187.87	308.7	879.22	900.84	1614.61	4933.61	4495.24
Financing of GFD—market borrowings	1.98	10.1	25.56	58.88	125.19	153.05	887.76	2840.5	3387.42
% of GFD financed from market borrowings	5.332615	13.42907	13.60515	19.07353	14.23876	16.9897	54.98294	57.57447	75.35571

Source of data RBI's Database on Indian Economy

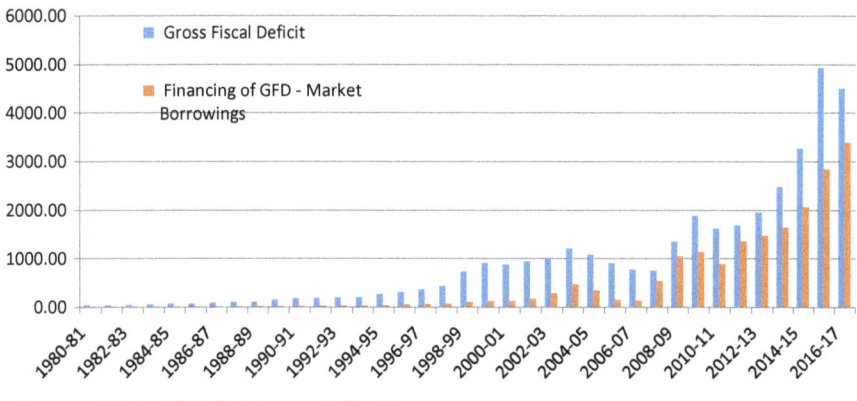

Source of Data: RBI's Database on Indian Economy

Fig. 5.2 States' gross fiscal deficit and its financing

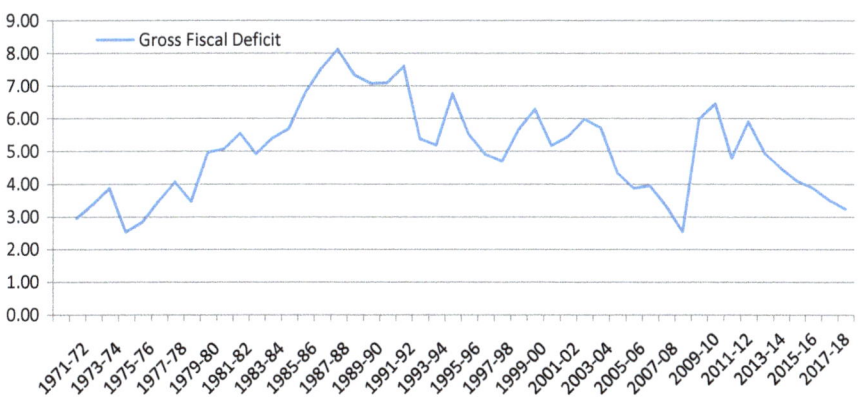

Source of Data: RBI's Database on Indian Economy

Fig. 5.3 Gross fiscal deficit of central government as percentage of GDP

Table 5.3 Market borrowings of central and state governments (combined) (Rupees billion)

Year	1980–81	1985–86	1990–91	1995–96	2000–01	2005–06	2010–11	2015–16	2016–17
Gross	32.04	71.78	115.58	467.83	1284.83	1817.47	5835.21	10335.93	11065.05
Net	28.11	60.74	105.7	327.21	866.67	1136.92	4147.97	7048.35	7393.41

Source of data RBI's Database on Indian Economy

The data on public issue of corporate debt shows that only one company issued public corporate debt in 2009 with issue size of Rs. 1500 crores, but in 2016 sixteen companies issued public corporate debt with the total issue size of

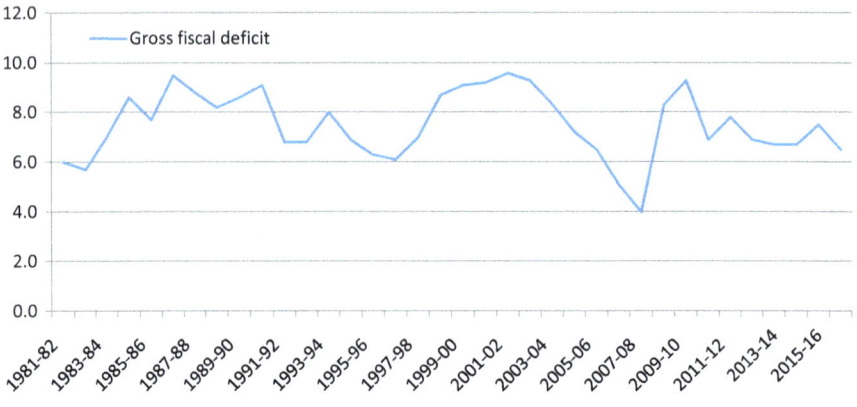

Source of Data: RBI's Database on Indian Economy

Fig. 5.4 Combined deficits of the central and state governments (as percentage to GDP)

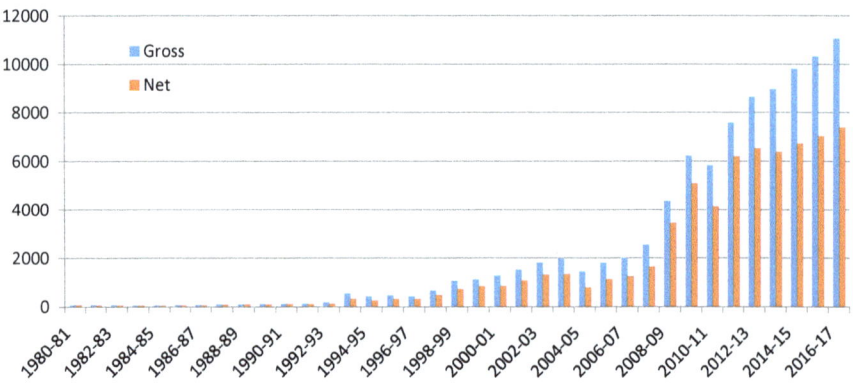

Source of Data: RBI's Database on Indian Economy

Fig. 5.5 Market borrowings of central and state governments (combined)

Rs. 29547.15 crores. Outstanding corporate debt in June 2010 was Rs. 735433.10 crores which has increased to Rs. 2586857.33 crores in September 2017. Total trading in corporate bonds in 2007–08 was Rs. 95889.706 crores which has increased to Rs. 1470662.51 in 2016–17. Source of data is Securities and Exchange Board of India (SEBI, n.d.). This data shows that the corporate bond market is expanding and moving toward a developed bond market.

Charts 5.1 and 5.2 (Khan 2016) show that the liquidity in both primary and secondary markets of Indian bond market is improving and bid–ask spread is decreasing.

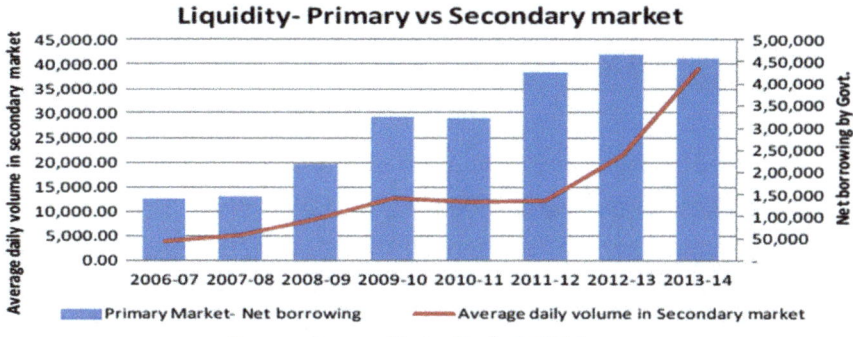

Chart 5.1 Liquidity in the primary and secondary markets

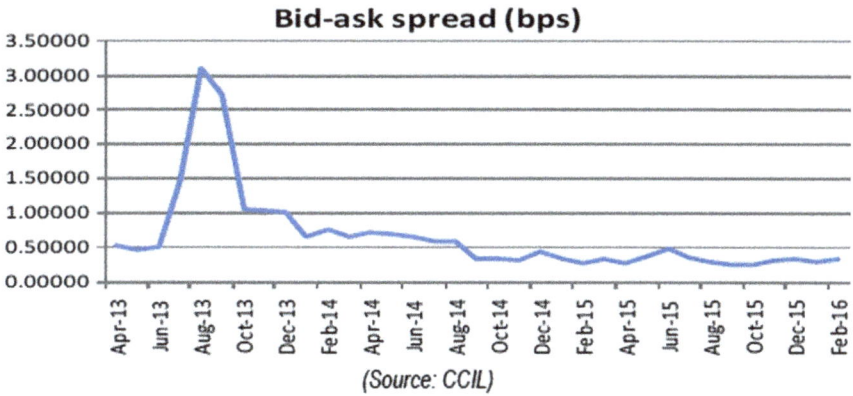

Chart 5.2 Bid–ask spread

5.5.3 Turnover of Government Securities Market (Face Value)

Turnover is one of the most important indicator of liquidity in the capital market. Due to improvement in the technology, screen-based trading, and other developments, there is significant increase in turnover in the government securities market in the last two decades. The turnover in the state government securities was very low as compared to central government securities. Share of central government securities has increased to more than 90% at the end of 2017. Table 5.4 shows that the turnover is also low in T-Bills. Total turnover has increased from 58 billion rupees in 1998 to 3858 billion rupees at the end of the year 2017. Figure 5.6 shows the increase in the turnover in the last few decades.

Table 5.4 Turnover in government securities market (face value)

Week ended	Outright transactions (Rs. Billion)			Total	Percentage of total		
	Central government dated securities	State government dated securities	T-Bills		Central government dated securities	State government dated securities	T-Bills
03-Apr-1998	37.42	0.41	20.57	58.40	64.08	0.70	35.22
25-Dec-1998	28.23	0.27	12.20	40.70	69.36	0.66	29.98
31-Dec-1999	98.45	2.17	17.66	118.28	83.23	1.83	14.93
29-Dec-2000	113.31	0.44	14.01	127.76	88.69	0.34	10.97
28-Dec-2001	150.11	2.58	26.32	179.01	83.86	1.44	14.70
27-Dec-2002	686.04	7.35	38.99	732.38	93.67	1.00	5.32
26-Dec-2003	420.13	3.74	33.52	457.39	91.85	0.82	7.33
31-Dec-2004	250.56	9.45	124.47	384.48	65.17	2.46	32.37
30-Dec-2005	107.77	3.55	28.04	139.36	77.33	2.55	20.12
29-Dec-2006	63.51	2.89	18.50	84.90	74.81	3.40	21.79
28-Dec-2007	368.31	6.13	61.47	435.91	84.49	1.41	14.10
26-Dec-2008	1115.41	24.34	60.39	1200.14	92.94	2.03	5.03
25-Dec-2009	863.71	29.35	111.13	1004.19	86.01	2.92	11.07
31-Dec-2010	660.64	12.78	57.12	730.54	90.43	1.75	7.82
30-Dec-2011	1494.45	12.16	129.65	1636.26	91.33	0.74	7.92
28-Dec-2012	1908.99	29.27	168.80	2107.05	90.60	1.39	8.01
27-Dec-2013	1256.70	25.53	279.46	1561.70	80.47	1.63	17.89
25-Dec-2015	2224.62	134.85	222.44	2581.91	86.16	5.22	8.62
30-Dec-2016	3446.51	245.72	692.78	4385.02	78.60	5.60	15.80
10-Nov-2017	3603.70	110.91	143.71	3858.32	93.40	2.87	3.72

Source of data RBI's Database on Indian Economy

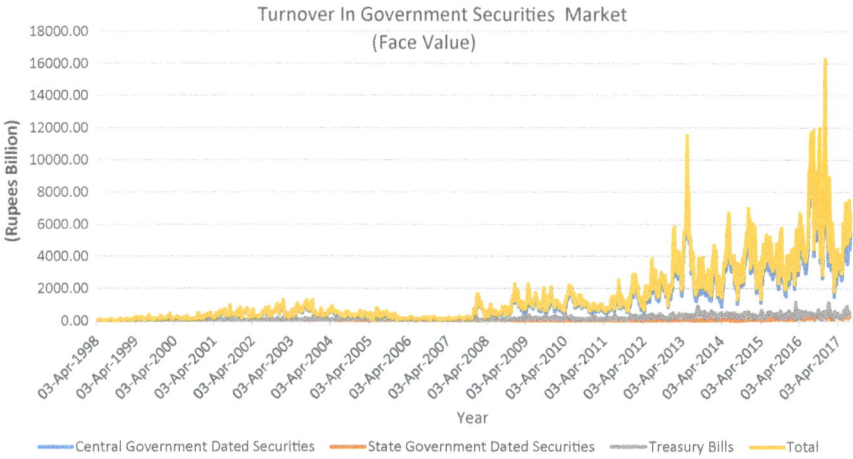

Fig. 5.6 Turnover in government securities market (face value)

5.5.4   Treasury Bills Outstanding

Treasury bills are the short-term source of finance for government of india. Government of india issues 14 days, 91 days, 182 days, 364 days, and cash management bills to raise short-term funds from the market. The bills outstanding are increased significantly over the years. 14 days, 91 days, 182 days, and 364 days T-Bills increased from 357.21, 231.43, 73.74, and 402.38, respectively, to 1127.99, 2045.75, 859.65, and 1403.91 billion rupees, respectively, from 2006 to 2017 and in percentage they have increased to 215.7778, 783.9606, 1065.785, and 248.9015%, respectively.

As it is clear from Table 5.5 that the total outstanding treasury in the year 2017 is more than Rs. 6000 billion. So the increase in T-Bills outstanding is very significant in the last one decade (Figs. 5.7 and 5.8).

5.5.5   Ownership Pattern of Government of India Dated Securities

One aspect of the Indian government securities market has not much changed in the last decade or so is the ownership pattern. Ownership in the Indian government securities market is dominated by the commercial banks, insurance companies, and RBI. Various participants in the Indian debt market are central government, state government, commercial banks, primary dealers, PSUs, corporates, and financial

Table 5.5 Government of india: treasury bills outstanding

Weed ended	14 day total (Rs. Billion)	14 day total (% increase from 2006)	91 day total (Rs. Billion)	91 day total (% increase from 2006)	182 day total (Rs. Billion)	182 day total (% increase from 2006)	364 day total (Rs. Billion)	364 day total (% increase from 2006)
12-May-06	357.21		231.43		73.74		402.38	
18-May-07	290.62	−18.6417	493.9	113.4123	192.49	161.0388	569.42	41.513
16-May-08	442.95	24.00269	470.48	103.2926	170.88	131.7331	589.25	46.44118
15-May-09	641.59	79.61143	800.03	245.6898	203.75	176.3087	494	22.76952
14-May-10	782.9	119.1708	735	217.5906	215	191.565	445.23	10.64914
13-May-11	727.08	103.5441	810.17	250.0713	272.51	269.5552	444.57	10.48511
18-May-12	735.6	105.9293	1473.10	536.5208	550.1	645.9995	993.8	146.9805
17-May-13	817.5	128.857	1164.15	403.0247	641.99	770.613	1304.89	224.293
16-May-14	675.59	89.12964	1588.28	586.2896	763.97	936.032	1408.15	249.9553
15-May-15	746.79	109.0619	1553.99	571.473	769.73	943.8432	1431.73	255.8154
13-May-16	1027.54	187.6571	1542.93	566.694	774.95	950.9222	1539.67	282.6408
12-May-17	1326.60	271.3782	1623.60	601.5512	872.24	1082.859	1412.50	251.0363
04-Nov-17	1127.99	215.7778	2045.75	783.9606	859.65	1065.785	1403.91	248.9015

Source of data RBI's Database on Indian Economy

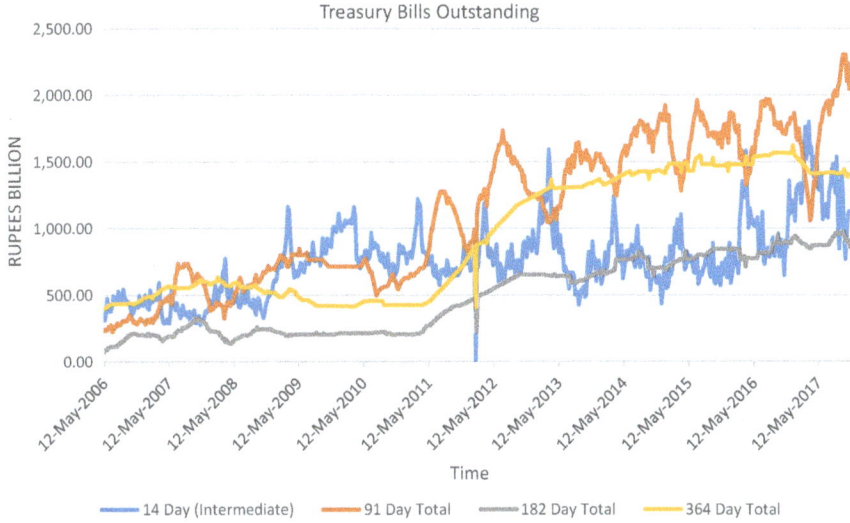

Source of Data: RBI's Database on Indian Economy

Fig. 5.7 Government of india: treasury bills outstanding

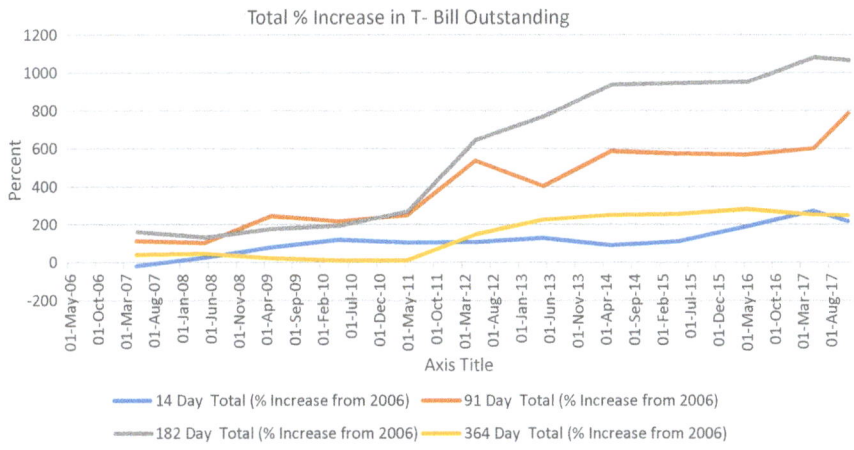

Source of Data: RBI's Database on Indian Economy

Fig. 5.8 Total percentage increase in T-Bill outstanding

institutions. Primary dealers are the market makers in the government securities market, underwrite the government securities, and help in the primary and secondary market operations. As it is clear from Fig. 5.9 that in March 2007

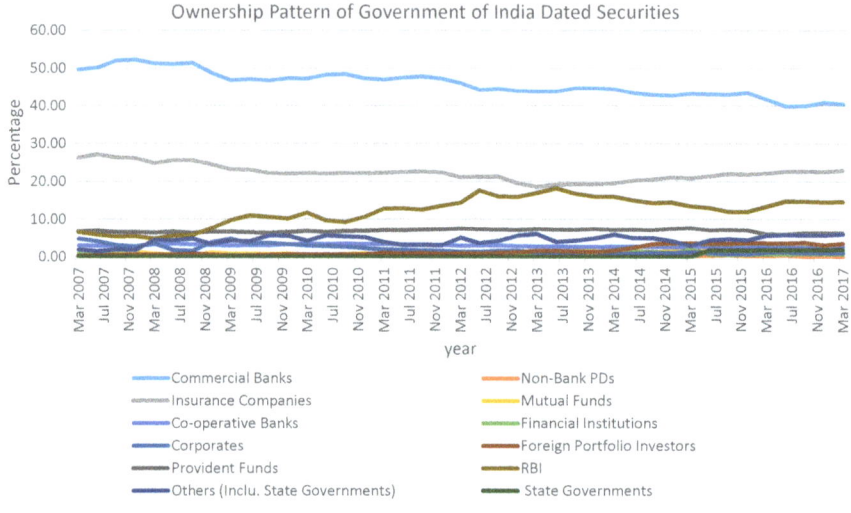

Source of Data: RBI's Database on Indian Economy

Fig. 5.9 Ownership pattern of government of india dated securities

commercial banks, insurance companies, and RBI owned 49.68, 26.19, and 6.51% of the government securities, respectively, which is changed in March 2017 to 40.46, 22.9, and 14.65%, respectively. So the domestic institutional investors are dominant players in Indian government securities market (Fig. 5.10).

Figure 5.8 shows the percentage ownership of foreign portfolio investors in government securities in India. As it is clear from Fig. 5.8 that the ownership has increased from 0.18% in March 2007 to 3.53 in March 2017. So the role of foreign portfolio investors is increasing in the Indian government securities market.

5.5.6 *Maturity Pattern of Government of India Rupee Loans*

Maturity pattern of government securities is also changing, and earlier the market was dominated by the securities with maturities more than 10 years. Table 5.6 shows how the maturity pattern has changed in the last four decades. The percentage of over 10-year bonds is decreasing and under 5-year bonds and between 5- and 10-years bonds are increasing. As it is clear from Fig. 5.11 that the over 10-year maturities still dominate but the bonds with maturities under 5-years and between 5 and 10 years are also increasing (Fig. 5.12).

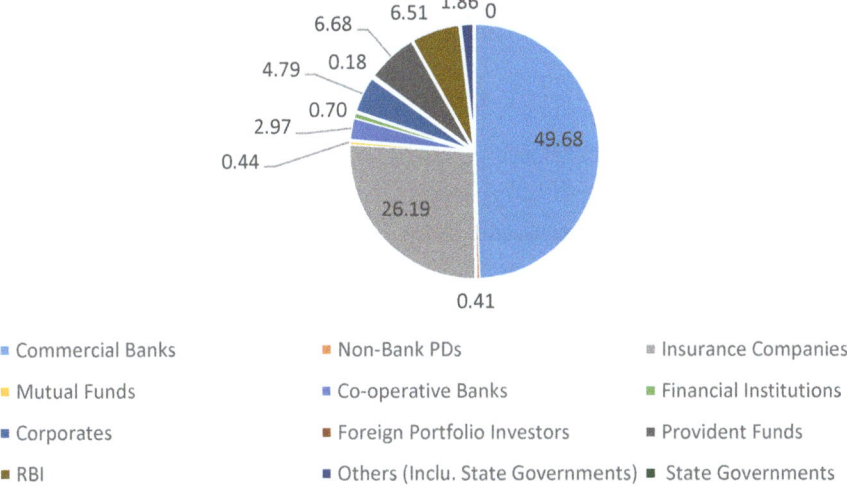

Ownership Pattern of Government of India Dated Securities March 2007

- Commercial Banks
- Non-Bank PDs
- Insurance Companies
- Mutual Funds
- Co-operative Banks
- Financial Institutions
- Corporates
- Foreign Portfolio Investors
- Provident Funds
- RBI
- Others (Inclu. State Governments)
- State Governments

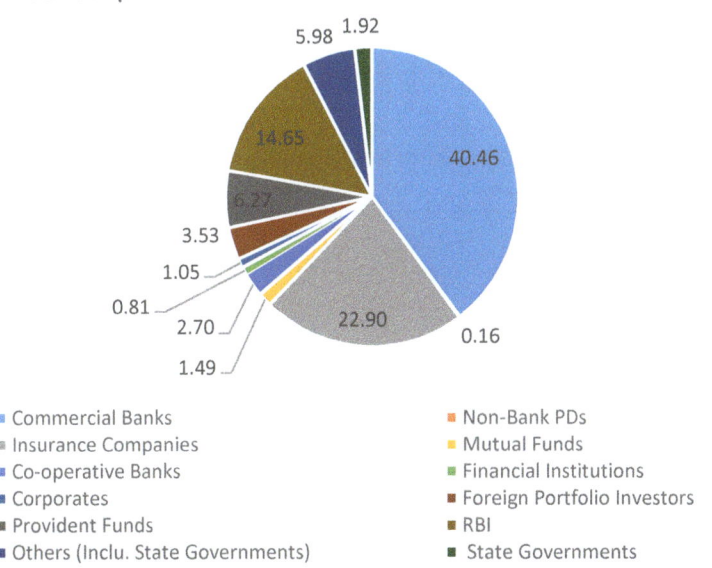

Ownership Pattern of Government of India Dated Securities March 2017

- Commercial Banks
- Non-Bank PDs
- Insurance Companies
- Mutual Funds
- Co-operative Banks
- Financial Institutions
- Corporates
- Foreign Portfolio Investors
- Provident Funds
- RBI
- Others (Inclu. State Governments)
- State Governments

Source of Data: RBI's Database on Indian Economy

Fig. 5.10 Ownership pattern of government of india dated securities

Table 5.6 Maturity pattern of government of india rupee loans

Year	Percent		
	Under 5 years	Between 5 and 10 years	Over 10 years
1977	15.1285	17.0343	65.0749
1987	9.79511	9.80765	80.3972
1997	40.9764	40.8326	18.191
2007	30.1336	29.8805	39.9859
2017	25.0397	33.2895	41.6694

Source of data RBI's database on Indian Economy rest

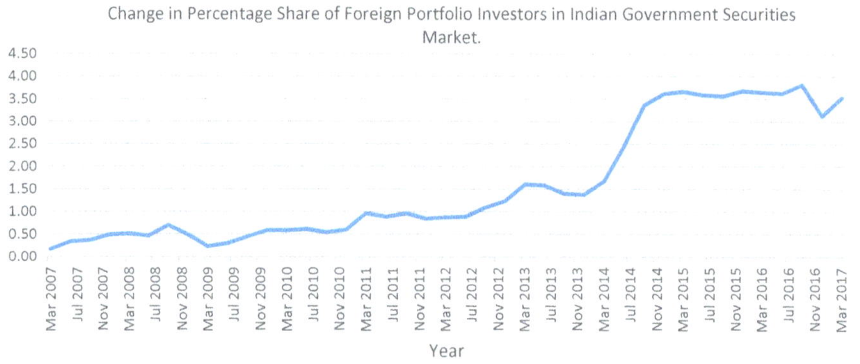

Change in Percentage Share of Foreign Portfolio Investors in Indian Government Securities Market.

Source of Data: RBI's Database on Indian Economy

Fig. 5.11 Change in percentage share of foreign portfolio investors in Indian government securities market

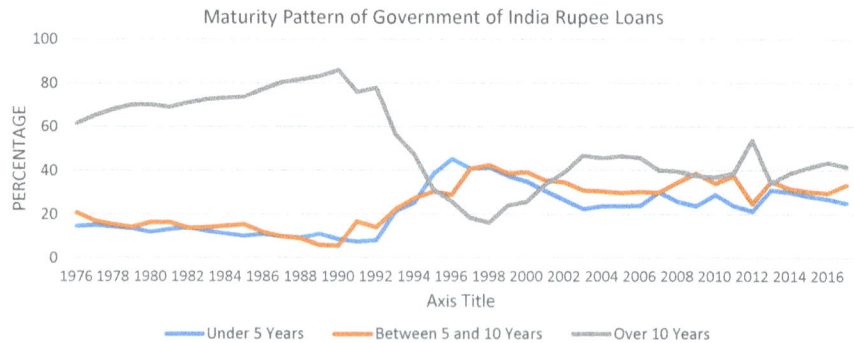

Source of Data: RBI's Database on Indian Economy

Fig. 5.12 Maturity pattern of government of india rupee loans

5.5.7   Indian 10-Year Government Bonds Yield

The benchmark yield of Indian 10-year government bonds is decreasing over the years. The 10-year government bond yield was 12.8% in the year 1994 and decreased to the lowest to 5.285 in the year 2004 but the standard deviation of the yield had been increased during the period. In the year 2009 and 2014, the yield increased to 6.681 and 8.662, respectively, but the standard deviation has decreased. Recently at the end of the year 2017, the yield was 7.34 and standard deviation was 0.6562. Standard deviation is a measure of risk. So it can be concluded that both risk and yield on the 10-year yield on government bonds have decrease in the last two decades. It is good sign for the Indian economy as government is paying less for borrowing from the market and risk in the Indian market has decreased. As it is also clear from Fig. 5.13 the 10-year government bond yield is fluctuating from 5 to 12% points. The source of Fig. 5.13 is tradingeconomics.com (Fig. 5.13; Table 5.7).

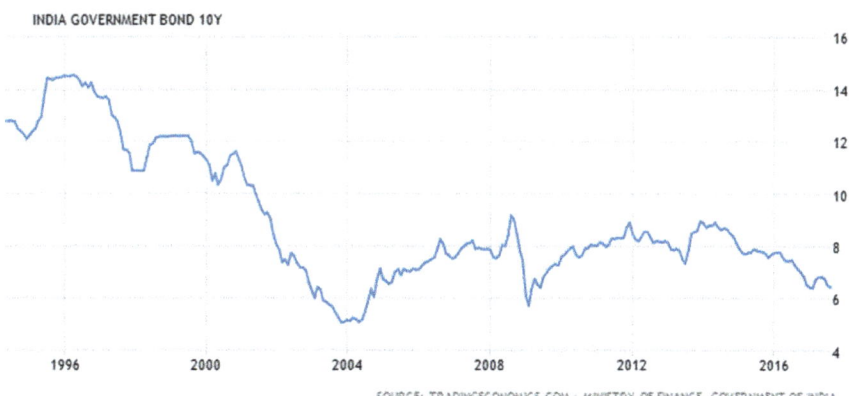

Fig. 5.13 Indian 10-year government bonds yield

Table 5.7 Indian 10-year government bonds yield

Year	01-05-1994	01-06-1999	01-06-2004	01-06-2009	01-06-2014	01-01-2018
10-year bond yield	12.8	12.22	5.285	6.681	8.662	7.34
St. Dev.		1.156724	2.409271	0.766376	0.499593	0.656205

5.6   Conclusion

Before 1990s, the Indian bond market was underdeveloped. In the last two decades, the Indian bond market has changed significantly. The market has become more liquid, transparent, and secure. Investor base is widening. With the innovation in the market infrastructure such as introduction of NDS-OM, CCIL, screen-based trading, network of primary dealers, allowing the government securities to be traded through stock exchanges, the volume in the secondary market has increased. With the growth of market borrowings by central and state government, the supply of the government securities has grown significantly in the last two decades. Beside the plain vanilla bonds the innovative bonds like zero-coupon bonds, inflation-indexed bonds, bonds with call and put options are also introduced in the Indian government securities market. RBI is no longer taking part in the primary market of government securities in India. Primary dealers act as underwriters in the market. RBI issues securities with a wide range of maturities. Market is still dominated by the Indian institutional investors but the role of foreign institutional investors is also increasing. With the improvement in liquidity, transparency, and security of investment in the government securities market, investors' base is widening. Retail investors are still reluctant to invest in the Indian government securities market. Ownership of these securities is still dominated by the commercial banks, insurance companies, and other institutional investors. The ownership of foreign portfolio investors has increased from 0.18% in March 2007 to 3.53 in March 2017. The maturity pattern is also changing. Over 10-year maturities still dominate but the bonds with maturities under 5 years and between 5- and 10-years are also increasing. There is significant increase in turnover in the government securities market in the last decade due to improvement in the technology, screen-based trading, and other developments. Benchmark 10-year government bond yield has decreased in the last decade. Amount of outstanding treasury bills has increased significantly over the years. In July 2017, outstanding treasury was more than Rs. 6000 billion. The Indian government securities market is moving from underdeveloped to the mature and developed debt market but the liquidity and depth in the market are still low as compare to the developed debt markets. Central government has issued securities worth of Rs. 15,000 crores every week in the 2017. The turnover in the market has also increased significantly. The impact of reforms found to be positive for all the dimensions, and these reforms have significant impact on the size and liquidity of the Indian bond market. The government securities market is more liquid than the corporate bond market and liquidity is still low as compared to the developed bond markets. Although liquidity is improving, in order to reduce the dependency on the banking sector, India needs further reforms, particularly in the corporate bond market. Even after the reforms the Indian government securities market cannot compete with the developed debt markets in other parts of the world. Challenges the Indian debt market still facing are low participation of the retail investors, low liquidity of some of the securities like securities issued by the state governments, which increases the cost of borrowings of the state governments, domination of the large institutional investors. To compete with the developed debt markets in the world, the liquidity, transparency, and depth of Indian government securities market required to be improved further. The study suggests that to improve liquidity in the Indian bond market the participation of retail investors need to be increased, access to the trading platform of government securities should be given to the small investors, government securities should be allowed to trade through the Demat account and tax benefits should be given to investors for investing in the government securities and corporate bonds.

References

Corsetti, G., Paolo, P. & Nouriel, R. (1998). What caused the Asian currency and financial crisis? Part I: A macroeconomic overview. *NBER Working Paper 6833*.

Endo, T. (2000). The development of corporate debt markets. Financial Markets Advisory Department, International Finance Corporation. *The World Bank Group*. www.worldbank.org

Fabella, R., & Madhur, S. (2003). Bond market development in East Asia: issues and challenges. Asian Development Bank *ERD Working Paper no. 1068*.

Goodfriend, M. (2005). *Based on the remarks made at Bank for International settlement seminar on "Corporate debt markets in India" held in Kunming, China*, November 17–18, 2005. www.bis.org

Harkansson, H. (1999). The role of a corporate bond market in an economy—and in avoiding crises. University of California, Berkeley. *Working Paper*. www.haas.berkely.edu

Herring, R. J., & Chatusripitak, N. (2000). The case of the missing market: The bond market and why it matters for financial development. *ADB Institute Working Paper Series No. 11*, July 2000.

Khan, H. R. (April, 2016). *Indian debt market 2020: The underpinnings & the path ahead*. Retrieved from https://www.rbi.org.in/scripts/BS_SpeechesView.aspx?Id=997

Luengnaruemitchai, P., & Ong L. L. (2005). An anatomy of corporate bond markets: growing pains and knowledge gains. *IMF Working Paper*. www.imf.org

Mohan, R. (August, 2004a). Speech at the 3rd India debt markets conference on debt markets in India—Issues and prospects, organized by Citigroup and Fitch Ratings India at Mumbai. Retrieved from https://www.rbi.org.in/Scripts/BS_SpeechesView.aspx?Id=180

Mohan, R. (May, 2004b). A decade of reforms in government securities market in India and the road ahead. RBI Bulletin, May 2004

Raghavan, S., & Sarwano, D. (2012). Development of the corporate bond market in India: An empirical and policy analysis. In *International Conference on Economics and Finance Research, IPEDR*, Vol. 32. www.ipedr.com

Rakshit, M. (2000). Crisis and recovery 1997–99 East Asia revisited. *Money & Finance*, No. 12.

RBI. (May, 2007). Report on government securities market. Retrieved from https://rbi.org.in/scripts/PublicationReportDetails.aspx?ID=503

Sabnavis, M. & Mehta, G. (2014). Indian Bond Market: Striking a Chord with Asian Peers. Professional Risk Opinion. Care Ratings. http://www.careratings.com

The International Capital Markets Association. (2013). Economic Importance of the Corporate Bond Markets. *ICMA*, Switzerland. https://www.icmagroup.org

Wells, S., & Schou-Zibel, L. (2008). India's bond market—Developments and challenges ahead. *Asian Development Bank Working Paper Series on Regional Economic Integration* No. 22. www.adb.org

RBI. (n.da). Brief history of public debt in India. Retrieved from https://rbi.org.in/history/Brief_Fun_PublicDebt.html#PublicDebt

RBI. (n.db). Database on Indian economy. Retrieved from https://dbie.rbi.org.in/DBIE/dbie.rbi?site=home

SEBI. (n.d.). *Corporate bonds*. Retrieved from https://www.sebi.gov.in/statistics/corporate-bonds.html

Chapter 6
Demand Forecasting of the Short-Lifecycle Dairy Products

Rahul S. Mor, Swatantra Kumar Jaiswal, Sarbjit Singh and Arvind Bhardwaj

Abstract Predictions of future market demands for dairy products are important determinants in developing marketing strategies and farm-production planning decisions. For business operations in dairy industry, the accuracy of the forecast is of crucial importance because of the volatile demand pattern, influenced by an environment of rapid and dynamic response. The current study aims to compare the forecasting models like moving average, regression, multiple regression, and the Holt–Winters model based on accuracy measures, applied to demand forecasting of a time series formed by a group of perishable dairy products in milk processing industry. Further, the metric analysis of various error-measuring techniques is also applied to select the least error-producing model for such products as a performance measure. Findings of the study will help dairy industry to achieve high order fill rate, good inventory control as well as high profits. However, the selection of these models depends upon the knowledge, availability of data, and context of forecasting.

Keywords Demand forecasting · Error measures · Dairy industry
Short-lifecycle products · Seasonality · Food processing

Nomenclature

F_t	Forecasted demand for the period 't'
Y_t	Actual demand for the period 't'
u_i	Level factor for the period 't'
v_i	Trend factor for the period 't'
S_i	Seasonal factor for the period 't'
α	Smoothing constant for demand or level

Electronic supplementary material The online version of this chapter (https://doi.org/10.1007/978-981-13-1334-9_6) contains supplementary material, which is available to authorized users.

R. S. Mor (✉) · S. K. Jaiswal · S. Singh · A. Bhardwaj
Department of Industrial & Production Engineering, National Institute of Technology, Jalandhar, Punjab, India
e-mail: dr.rahulmor@gmail.com

© Springer Nature Singapore Pte Ltd. 2019   87
H. Chahal et al. (eds.), *Understanding the Role of Business Analytics*,
https://doi.org/10.1007/978-981-13-1334-9_6

β Smoothing constant for trend
γ Smoothing constant for seasonality
MAD   Mean absolute deviation
MSE   Mean square error
RMSE   Root-mean-square error
MAPE   Mean absolute percentage error
M.A.   Moving average
H-W   Holt and Winters

6.1 Introduction

India is the largest producer of milk in the world, and it is also the largest consumer of milk, consuming almost its whole milk production. The dairy industry has been a significant contributor to the gross domestic product in India, and its value of output has grown significantly where the milk is processed and marketed by 170 milk producers' cooperative unions, which federate into 22 state cooperative milk marketing federations. The organized sector still remains a minor stakeholder and handles about 20% of the milk, whereas the unorganized sector of *dudhiyas* and *mithaiwallas* still controls about 80% of the industry. There is a wide range of dairy products in India such as milk (full cream, toned, double-toned, cow milk, standard milk), curd, lassi (plain lassi, sweet lassi, spicy lassi), paneer, ghee, kheer (simple kheer, kesar kheer), ice cream, butter, Pio milk 200-ml bottle, panjiri, khoya, processed cheese, kaju pinni, jal jeera. The dairy industry needs key development in the effectiveness of supply chain so as to meet the high-quality, reliability, and safety standards of the export markets (Bhardwaj et al. 2016; Mor et al. 2018a, b, c, d). Forecasting means estimating future event by casting the forward past data where past data is systematically combined in a predetermined way to get the future estimate. In other words, forecasting is the process of making predictions of the future based on the past and present data and most commonly by analysis of trends or by analysis of seasonality. Forecasting may be used for long-range planning, intermediate-range planning, and short-term control (Kaloxylos et al. 2013). Analysis of data requires the analyst to identify the underlying behavior of the series. This can often be accomplished by merely plotting the data and visually examining the plot at a specified interval of time. One or more patterns might appear like trends, seasonal variations, cycles, and variations around an average (Yuan and Cai 2008). In addition, there can be various terms such as trend, seasonality, cyclic variation, random or irregular variations. Both qualitative and quantitative methods are used for forecasting (Mor et al. 2018e). The *quantitative methods* of forecasting are based on opinions, intuition, or personal experiences and are subjective in nature. They do not rely on rigorous mathematical computations and permit inclusion of human factor and are used for long- or intermediate-range decision of the new product (Leat and Giha 2008). Whereas quantitative methods use the past data to forecast through some statistical tools and are usually applied to short or intermediate range decisions. These use time-series and casual methods for forecasting.

6.2 Literature Review

Mishra et al. (2016) observed the increase in expected number of items requested by customers due to that with the sales forecasting, and it also affects the production in order to produce goods. Ping (2016) aimed to predict their future handling capacity based on the regression analysis model of the upper Yangtze River port through forecasting. Sugiarto et al. (2016) used the sales and distribution module to simplify the process of selling to customers in accordance with the interests of customers for goods and services, make it easy to check the sale of goods and delivery of goods, and facilitate the collection of customers. Tratar and Strmčnik (2016) said that if an error occurs caused by the production of demand forecasting, it will affect the process of the sale of goods to customers, and demand forecasting can also be used for the avoidance of excess or shortage of supply of goods in the warehouse. Gupta (2015) and Sarno and Herdiyanti (2010) worked on forecasting the packaged food product demand using mathematical programming and found demand forecasting as most common phenomenon observed in the industry. Harsoor and Patil (2015) made an attempt by understanding the retail store business's driving factors by analyzing the sales data of Walmart store that is geographically located at various locations and the forecast of sales. Hassan et al. (2015) proposed the procedure for new product sales forecasting which guides the calculation of new product sales forecasts based on accusation, evaluation, and choice of subjective forecasts provided by executives and sales team for new products which do not have any historical data. Mor et al. (2015, 2017) concluded that the sustainable agri-food supply chains can be achieved through innovation, supply chain collaboration, elimination of uncertainties. Zhou et al. (2015) found that a reliable inventory prediction can avoid product overstock and reduces the maintenance cost, and proposed two-step dynamic forecasting model which is capable to capture the characteristics of stock out time series. García et al. (2014) emphasized on the strategic connotations to do with packaging design being one of the supports of competitive advantages in the supply chain management.

Patushi and Kume (2014) suggested the cluster development as a way to increase competitiveness in business and a way for the effective use of the potential of the dairy processing industry through policy guidance and their management to meet the challenges, focusing primarily the region of Tirana. Veiga et al. (2014) aimed to compare the performances between ARIMA and Holt–Winters (H-W) models for the prediction of a time series formed by a group of perishable dairy products. Weber et al. (2014) dealt with the interactions of prices concerning different marketing levels in the German dairy sector focusing on whole milk powder. Assis et al. (2013) proposed a traffic characterization using two-dimensional flow analysis for modeling the behavior traffic pattern, here called digital signature of network segment using flow analysis. Spicka (2013) presented an in-depth view on the competitive environment of the Czech dairy industry through Porter's five forces analysis and concluded that the vertical business relationships within dairy supply chain can be considered as the weakness of the Czech dairy industry. Amorim et al. (2013) presented the trade-off by developing risk-averse production planning models and suggested that it is possible to reduce the percentage of expired products that reach the end of their shelf lives by using the risk-averse models. Taylor (2011) found that efficient supply chain management relies on accurate demand forecasting. Jraisat et al. (2013) stated that a company can sustain in business with the sale, but it must have a strong sales function and also be able to distribute goods to customers quickly and efficiently. Ghosh (2008) used the univariate time-series methods like multiplicative, seasonal, autoregressive, integrated, moving average, and Holt–Winters multiplicative exponential smoothing to forecast the monthly peak demand of electricity in India. Dhahri and Chabchoub (2007) applied the ARIMA model for forecasting the sugarcane productions in India with annual data. Vaida (2008) dealt with the theoretical aspects of the market demand method selection criteria and their application in developing the Lithuanian furniture demand forecast.

6.2.1   Research Gaps

Forecasting is pretty basic practice in short-lifecycle products, and many of the dairy industries have not adopted any advanced forecasting techniques. The day-to-day demand in these industries is recorded on the basis of the information given by the customers or retailers through telephonic conversation or other informal methods. This method is simple yet sometimes proves efficient. But due to the highly perishable nature of dairy items, a perfect forecasting technique is must which leads to efficient production as well as the inventory management. This leads to low order fill rate and wastages due to which overburden on the milk processing plant is caused to meet the uncertain demand. All these reasons motivated authors to design a research problem in the current study.

6.3   Methodology

This study started with an aim to identify various forecasting methods. The second aim of the study is to propose a forecasting framework for short-lifecycle dairy products. After the comprehensive literature review, pilot study, and discussion with top-level managers of case industry, the problem is identified. Data is collected from milk processing units located at northern India. The demand pattern analysis is carried out based on collected data, and suitable forecasting models have been nominated. Seasonality factor is also considered while estimating the demand, and the hypothesis is framed. Findings of the study have been carried out while considering the accuracy measures also as follows.

6.3.1 Data Collection

A milk processing unit located in northern India was selected to execute the demand forecasting of its perishable products in the current study. In order to forecast the demand of the case industry, four-year data (from April 2013 to March 2017) was collected through personal visits to various departments such as marketing, sales, and procurement. The collected data was then converted into an excel spreadsheet.

6.3.2 Demand Pattern Analysis

The analysis of data requires the analyst to identify the underlying behavior of the series. This can often be accomplished by merely plotting the data in MS Excel and visually examining the plot at a specified interval of time. One or more patterns might appear such as trends, seasonal variations, cycles, and variations around an average. Seasonality may refer to regular annual variations where the variations in time-series data repeat upward or downward movements. The term seasonal variation is also applied to daily, weekly, monthly, and other regularly recurring patterns in data. Seasonality in a time series is expressed in terms of the amount that actual values deviate from the average value of a series. If the series tends to vary around an average value, then seasonality is expressed in terms of that average (or a moving average), but if the trend is present, seasonality is expressed in terms of the trend value. There are two different models of seasonality as additive and multiplicative. In the additive model, seasonality is expressed as a quantity (e.g., 20 units), which is added or subtracted from the series average in order to incorporate seasonality. In the multiplicative model, seasonality is expressed as a percentage of the average (or trend) amount (e.g., 1.10), which is then used to multiply the value of a series to incorporate seasonality. Demand pattern found in various dairy products of case milk processing industry is mentioned below:

    i. **Milk**: During forecasting analysis, it was found that the milk demand follow seasonal pattern and no trend was observed. So, it has to be taken into analysis as it can help in minimizing the error. Milk demand pattern is shown in Fig. 6.1.
   ii. **Curd**: As shown in Fig. 6.2, curd is also following monthly seasonal demand pattern.
  iii. **Paneer**: The demand pattern of paneer is shown in Fig. 6.3.
  iv. **Plain Lassi**: The demand pattern of plain lassi is shown in Fig. 6.4.
   v. **Kheer**: The demand pattern of kheer is shown in Fig. 6.5.
  vi. **Ice Cream**: The demand pattern of ice cream is shown in Fig. 6.6.
 vii. **Table Butter**: The demand pattern of table butter is shown in Fig. 6.7.
viii. **Pio 200-ml Bottle**: The demand pattern of Pio 200-ml bottle is shown in Fig. 6.8.
  ix. **Ghee**: The demand pattern of ghee is shown in Fig. 6.9.

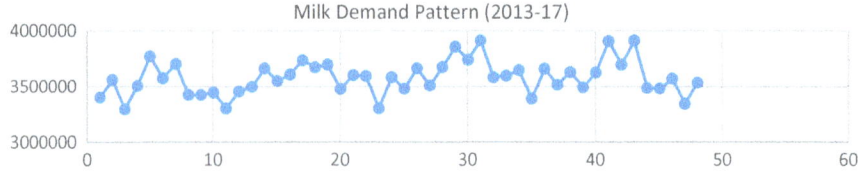

Fig. 6.1 Demand pattern for milk

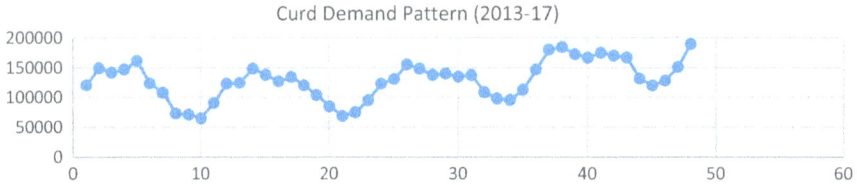

Fig. 6.2 Demand pattern for curd

Fig. 6.3 Demand pattern for paneer

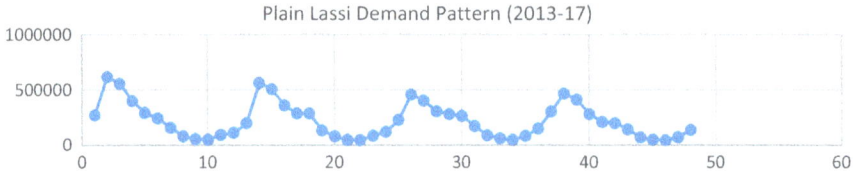

Fig. 6.4 Demand pattern for plain lassi

Fig. 6.5 Demand pattern for kheer

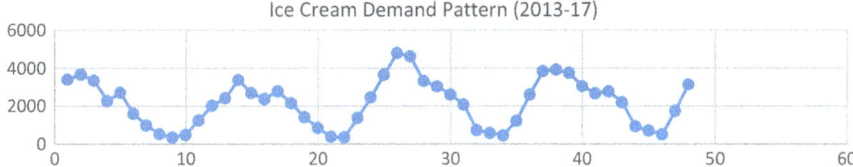

Fig. 6.6 Demand pattern for ice cream

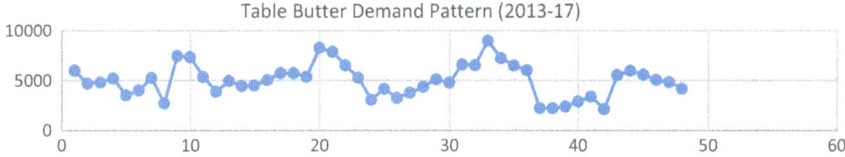

Fig. 6.7 Demand pattern for table butter

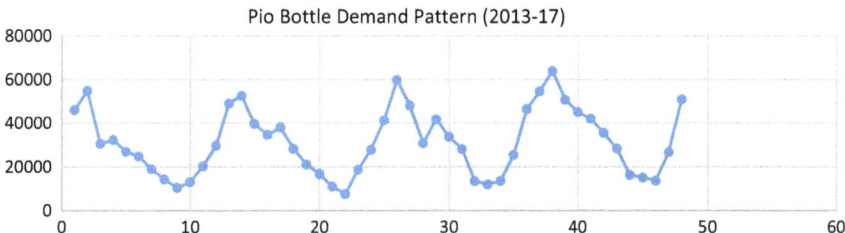

Fig. 6.8 Demand pattern for Pio bottle

Fig. 6.9 Demand pattern for ghee

6.4 Analysis, Result, and Discussions

The data collected from the select dairy industry has been analyzed by using MS Excel, and the results obtained from the forecasting methods' analysis are mentioned below.

6.4.1 Forecasting Results

6.4.1.1 Forecasting Results for Milk

For demand analysis of milk, different forecasting techniques were used and four different types of errors were calculated, i.e., MAPE, MAD, MSE, and RMSE (Table 6.1).

From Table 6.1, the method with minimum error will be chosen for the forecasting of milk and demand analysis (Fig. 6.10).

The errors calculated for multiple regression technique are minimum. Thus, multiple regression technique is best for the forecasting of milk.

Table 6.1 Milk demand analysis

S. No.	Forecasting methods	MAPE	MAD	MSE	RMSE			
1	MA5	3.820019	136844.9	26,637,481,767	163,209.9			
2	MA6	4.012007	143,867.3	30,353,610,176	174,222.9			
3	MA7	3.87788	139,026.7	28,730,615,087	169,501.1			
4	MA8	3.839402	138,103.3	28,628,503,966	169,199.6			
5	MA9	3.648302	131,416.6	27,807,111,090	166,754.6			
6	H-W	1.552632	55,706.5	5,411,244,645	73,561.16	Optimum value		
						α	β	γ
						0.04	0.95	0.2
7	Linear regression	3.2835	117,644.3	21,625,623,286	147,056.5			
8	Multiple regression	1.30297	46,632.8	2,928,077,122	54,111.71			
9	Multiple regression 2	3.5531	125,871.5	18,889,692,267	137,439.8			

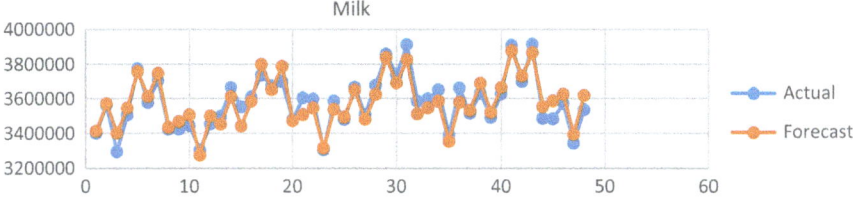

Fig. 6.10 Comparison of actual demand versus forecast of milk

6.4.1.2 Forecasting Results for Curd

For demand analysis of curd, different forecasting techniques were used and four different types of errors were calculated, i.e., MAPE, MAD, MSE, and RMSE (Table 6.2).

From Table 6.2, the method with minimum error will be chosen for the forecasting of curd and demand analysis (Fig. 6.11).

The errors calculated for Holt–Winters technique are minimum. Thus, Holt–Winters technique is best for the forecasting of curd.

Table 6.2 Curd demand analysis

S. No.	Forecasting methods	MAPE	MAD	MSE	RMSE			
1	MA5	24.91038	28,190.63	1,165,788,899	34,143.65			
2	MA6	26.85498	30,236.56	1,265,533,987	35,574.34			
3	MA7	27.81183	31,207.15	1,303,164,214	36,099.37			
4	MA8	26.9243	31,021.88	1,229,869,877	35,069.5			
5	MA9	25.41938	30,251.85	1,113,538,210	33,369.72			
6	H-W	5.8111	7331.148	96,106,553.08	9803.395	Optimum value		
						α	β	γ
						0.1	0.95	0.95
7	Linear regression	21.7303	24,109.44	834,042,945.9	28,879.8			
8	Multiple regression	7.212	9278.54	134,994,511.6	11,618.71			
9	Multiple regression 2	18.157	28,624.71	918,927,754.6	30,313.82			

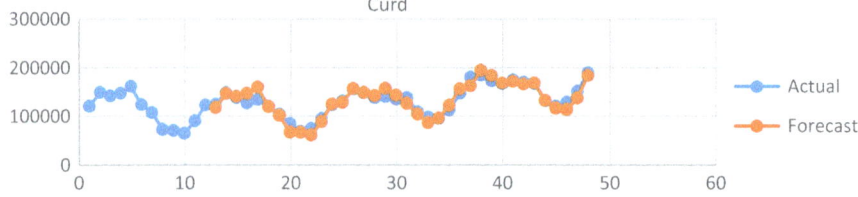

Fig. 6.11 Comparison of actual demand versus forecast of curd

6.4.1.3 Forecasting Results for Paneer

For demand analysis of paneer, different forecasting techniques were used and four different types of errors were calculated, i.e., MAPE, MAD, MSE, and RMSE (Table 6.3).

From Table 6.3, the method with minimum error will be chosen for the forecasting of paneer and demand analysis (Fig. 6.12).

Table 6.3 Paneer demand analysis

S. No.	Forecasting methods	MAPE	MAD	MSE	RMSE			
1	MA5	14.17066	2574.06	10,537,647.14	3246.174			
2	MA6	13.54214	2462.133	10,238,182.52	3199.716			
3	MA7	11.65128	2195.213	8,952,866.56	2992.134			
4	MA8	11.42924	2210.491	8,956,561.717	2992.752			
5	MA9	11.29246	2241.011	9,256,606.814	3042.467			
6	H-W	10.645	2049.809	63,332,577	2516.461	Optimum value		
						α	β	γ
						0.2	0	0.95
7	Linear regression	12.354	2206.103	8,248,411	2872			
8	Multiple regression	9.52	1727.61	4,677,638	2162.79			
9	Multiple regression 2	10.12	2310.25	9,525,315	3086.31			

Fig. 6.12 Comparison of actual demand versus forecast of paneer

The errors calculated for multiple regression technique are minimum. Thus, multiple regression technique is best for the forecasting of paneer.

6.4.1.4 Forecasting Results for Plain Lassi

For demand analysis of plain lassi, different forecasting techniques were used and four different types of errors were calculated, i.e., MAPE, MAD, MSE, and RMSE (Table 6.4).

From Table 6.4, the method with minimum error will be chosen for the forecasting of plain lassi and demand analysis (Fig. 6.13).

The errors calculated for multiple regression technique are minimum. Thus, multiple regression technique is best for the forecasting of plain lassi.

Table 6.4 Plain lassi demand analysis

S. No.	Forecasting methods	MAPE	MAD	MSE	RMSE			
1	MA5	116.3831	146,794.3	31,755,752,075	178201.4			
2	MA6	133.8051	153,265.9	35,339,066,828	187,986.9			
3	MA7	147.8859	156,355.3	36,523,227,245	191,110.5			
4	MA8	150.9534	156,293.5	35,149,012,148	187,480.7			
5	MA9	142.1247	152,584.6	32,271,211,772	179,641.9			
6	H-W	12.56711	29,733.87	2,872,405,609	53,594.83	Optimum value		
						α	β	γ
						0.9	0	0.05
7	Linear regression	103.36	124,624.5	22,558,506,710	150,194.9			
8	Multiple regression	15.78	26,173.17	996,408,394.4	31,565.94			
9	Multiple regression 2	25.97	39,065.02	2,033,333,297	45,092.5			

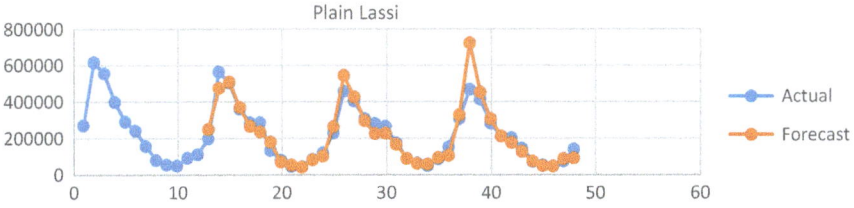

Fig. 6.13 Comparison of actual demand versus forecast of plain lassi

6.4.1.5 Forecasting Results for Kheer

For demand analysis of Kheer, different forecasting techniques were used and four different types of errors were calculated, i.e., MAPE, MAD, MSE, and RMSE (Table 6.5).

From Table 6.5, the method with minimum error will be chosen for the forecasting of kheer and demand analysis (Fig. 6.14).

The errors calculated for multiple regression technique are minimum. Thus, multiple regression technique is best for the forecasting of kheer.

Table 6.5 Kheer demand analysis

S. No.	Forecasting methods	MAPE	MAD	MSE	RMSE			
1	MA5	34.58572	2925.877	12,087,493.94	3476.707			
2	MA6	36.91362	3129.921	12,860,550.39	3586.161			
3	MA7	38.4764	3268.201	13,429,585.37	3664.64			
4	MA8	38.49255	3293.778	13,618,657.23	3690.346			
5	MA9	36.64657	3234.154	13,109,667.8	3620.728			
6	H-W	9.628	1069.402	1,950,975	1396.773	Optimum value		
						α	β	γ
						0.7	0	0.8
7	Linear regression	30.119	2401.91	8,206,995	2864.78			
8	Multiple regression	8.87	817.31	1,081,155	1039.786			
9	Multiple regression 2	22.14	2159.16	5,727,059	2393.12			

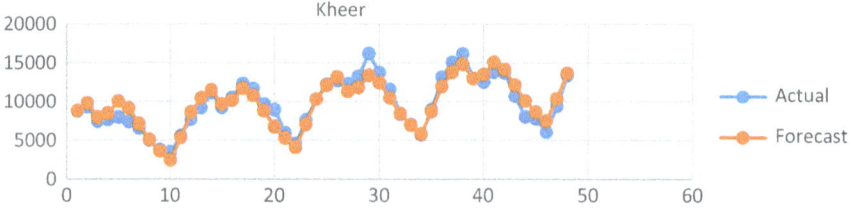

Fig. 6.14 Comparison of actual demand versus forecast of kheer

6.4.1.6   Forecasting Results for Ice Cream

For demand analysis of ice cream, different forecasting techniques were used and four different types of errors were calculated, i.e., MAPE, MAD, MSE, and RMSE (Table 6.6).

From Table 6.6, the method with minimum error will be chosen for the forecasting of ice cream and demand analysis (Fig. 6.15).

The errors calculated for multiple regression technique are minimum. Thus, multiple regression technique is best for the forecasting of ice cream.

Table 6.6 Ice cream demand analysis

S. No.	Forecasting methods	MAPE	MAD	MSE	RMSE			
1	MA5	114.0027	1295.813	2,249,901.515	1499.967			
2	MA6	126.2364	1353.883	2,479,202.677	1574.548			
3	MA7	135.0631	1365.076	2,578,585.739	1605.798			
4	MA8	135.5148	1362.211	2,525,711.811	1589.249			
5	MA9	124.5884	1317.24	2,338,875.246	1529.338			
6	H-W	18.81674	412.5208	303,085.3	550.5318	Optimum value		
						α	β	γ
						0.05	0.2	0.5
7	Linear regression	105.76	1043.95	1,501,256	1225.25			
8	Multiple regression	14.92	240.95	103,270.5	321.35			
9	Multiple regression 2	22.2	307.54	140,380.8	374.67			

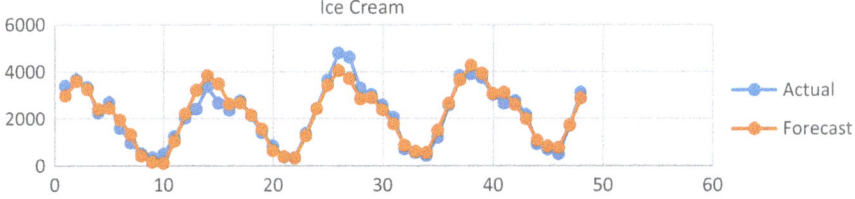

Fig. 6.15 Comparison of actual demand versus forecast of ice cream

6.4.1.7   Forecasting Results for Table Butter

For demand analysis of table butter, different forecasting techniques were used and four different types of errors were calculated, i.e., MAPE, MAD, MSE, and RMSE (Table 6.7).

From Table 6.7, the method with minimum error will be chosen for the forecasting of table butter and demand analysis (Fig. 6.16).

The errors calculated for the multiple regression technique are minimum. Thus, multiple regression technique is best for the forecasting of table butter.

Table 6.7 Table butter demand analysis

S. No.	Forecasting methods	MAPE	MAD	MSE	RMSE			
1	MA5	36.81538	1521.722	3,699,990.651	1923.536			
2	MA6	39.07002	1591.702	4,012,062.866	2003.013			
3	MA7	41.28587	1670.318	4,200,824.363	2049.591			
4	MA8	40.25224	1640.179	4,139,873.408	2034.668			
5	MA9	39.44391	1560.724	3,897,315.163	1974.162			
6	H-W	32.16	1367.779	2,762,688	1662.123	Optimum value		
						α	β	γ
						0.01	0.95	0.95
7	Linear regression	30.636	1267.076	2,533,924	1591.83			
8	Multiple regression	20.457	850.833	1,143,088	1069.15			
9	Multiple regression 2	70.05	2099.039	5,365,511	2316.35			

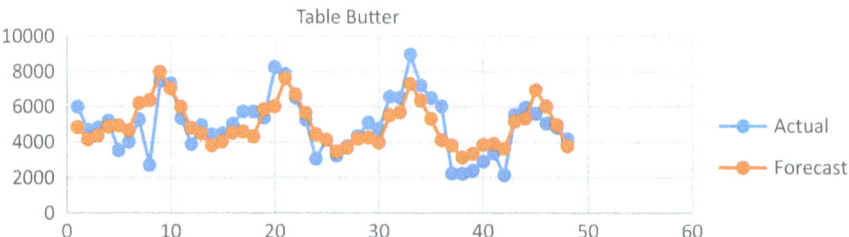

Fig. 6.16 Comparison of actual demand versus forecast of table butter

6.4.1.8 Forecasting Results for Pio bottle

For demand analysis of Pio bottle, different forecasting techniques were used and four different types of errors were calculated, i.e., MAPE, MAD, MSE, and RMSE (Table 6.8).

From Table 6.8, the method with minimum error will be chosen for the forecasting of Pio bottle and demand analysis (Fig. 6.17).

The errors calculated for the multiple regression 2 technique are minimum. Thus, multiple regression 2 technique is best for the forecasting of Pio bottle.

Table 6.8 Pio bottle demand analysis

S. No.	Forecasting methods	MAPE	MAD	MSE	RMSE			
1	MA5	63.70856	15,403.67	324,812,176.7	18,022.55			
2	MA6	68.46004	15,852.38	356,790,213.5	18,888.89			
3	MA7	72.72545	16,225.98	372,384,886.5	19,297.28			
4	MA8	74.81113	16,404.69	371,399,301.6	19,271.72			
5	MA9	71.97338	16,101.73	349,555,832.6	18,696.41			
6	H-W	13.34916	3963.806	28,279,949	5317.889	Optimum value		
						α	β	γ
						0.1	0.05	0.5
7	Linear Regression	58.06	12,567.84	285,775,316.3	16,904.89			
8	Multiple Regression	11.96	2893.32	13,251,240	3640.225			
9	Multiple Regression 2	10.675	3312.667	19,186,587	4380.25			

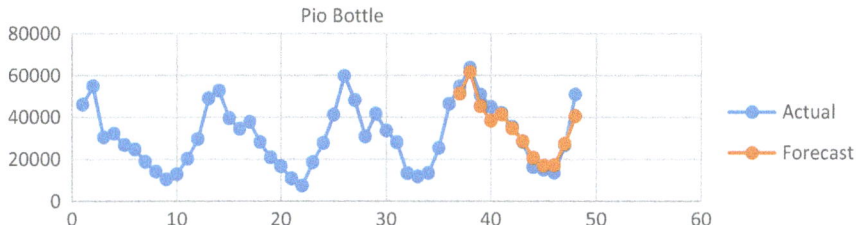

Fig. 6.17 Comparison of actual demand versus forecast of Pio bottle

6.4.1.9   Forecasting Results for Ghee

For demand analysis of ghee, different forecasting techniques were used and four different types of errors were calculated, i.e., MAPE, MAD, MSE, and RMSE (Table 6.9).

From Table 6.9, the method with minimum error will be chosen for the forecasting of ghee and demand analysis (Fig. 6.18).

The errors calculated for the multiple regression technique are minimum. Thus, multiple regression technique is best for the forecasting of ghee.

Table 6.9 Ghee demand analysis

S. No.	Forecasting methods	MAPE	MAD	MSE	RMSE	Optimum value		
1	MA5	53.35496	9831.307	133,235,331	11,542.76			
2	MA6	52.25332	10,050.51	140,643,879.7	11,859.34			
3	MA7	53.21595	10,251.76	144,702,910.2	12,029.25			
4	MA8	43.84376	9669.556	134,086,428.8	11,579.57			
5	MA9	41.48676	9048.746	124,128,291.7	11,141.29			
6	H-W	33.12453	6828.214	83,040,822.35	9112.674	Optimum value		
						α	β	γ
						0.1	0	0.95
7	Linear regression	45.44	7713.63	91,728,710	9577.512			
8	Multiple regression	32.202	5177.805	41,942,027.5	6476.266			
9	Multiple regression 2	37.559	6879.486	64,677,426	8042.228			

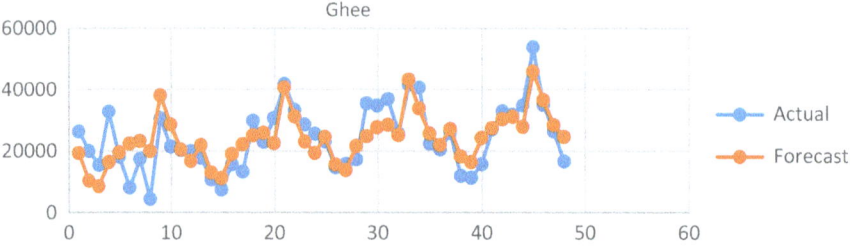

Fig. 6.18 Comparison of actual demand versus forecast of ghee

6.4.1.10   Forecasting Results for Milk: Daily Demand

For demand analysis of milk (daily demand basis), different forecasting techniques were used and four different types of errors were calculated, i.e., MAPE, MAD, MSE, and RMSE (Table 6.10).

From Table 6.10, the method with minimum error will be chosen for the forecasting of milk and demand analysis (Fig. 6.19).

The errors calculated for the Holt–Winters technique are minimum. Thus, H-W technique is best for the forecasting of milk (daily demand).

Table 6.10 Milk (based on daily demand) demand analysis

S. No.	Forecasting methods	MAPE	MAD	MSE	RMSE			
1	H-W	2.994861	3134.495	52,411,611.35	7239.586	Optimum value		
						α	β	γ
						0.2	0	0.01
2	Linear regression	4.178335	4608.72	60,900,472.83	7803.876			
3	Multiple regression	4.070971	4484.643	59,452,294.76	7710.531			
4	Multiple regression 2	5.675612	6319.01	86,816,156.89	9317.519			

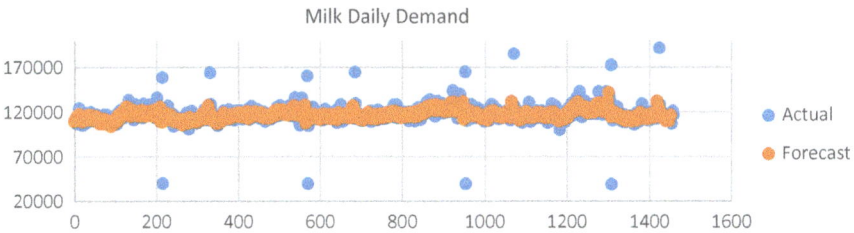

Fig. 6.19 Comparison of actual demand versus forecast of milk (based on daily demand)

6.4.2   Hypothesis

Following hypothesis is framed for linear regression case:
  H_0: *Whether the coefficients values are zero or not.*

1. After analyzing the regression result, we can conclude that R-squared value (coefficient of determination) is low. It is not near to 1.
2. As adjusted R-squared value is less than R-squared value, which shows that our model run is good and is in favor of results obtained, as shown below:

Residuals				
Min	1Q	Median	3Q	Max
−26.172	−8.157	2.163	8.346	22.197

Coefficients				
	Estimate	Std. error	t value	Pr(>|t|)
(Intercept)	7.428e−01	7.742e+00	0.096	0.92398
Actual	1.836e−04	5.811e−05	3.160	0.00279

Signif. codes: 0 Ô***Õ 0.001 Ô**Õ 0.01 Ô*Õ 0.05 Ô.Õ 0.1 Ô Õ 1

Residual standard error: 12.83 on 46 degrees of freedom

Multiple R-squared: 0.1784, adjusted R-squared: 0.1605

F-statistic: 9.987 on 1 and 46 DF, p value: 0.002788

P value is less than 0.05, and hence, null hypothesis is accepted.

Thus, demand forecasting assists other subsidiary departments of the industry such as finance, human resources, marketing. Food products show frequent changes in demand for different seasons, and thus, it is very necessary to forecast the demand of these products precisely. The present study analyzes the results obtained from the above-mentioned methods of forecasting for a dairy industry, thereby selecting an effective forecasting method for better results. However, the selection of these models depends upon the knowledge, availability of data, and context of forecasting.

6.5   Conclusions

Seasonal products are such a short-life span type of products in which the tolerance of excess stocks is lesser than for non-perishables. Demand forecasting is essential for perishable products like dairy industry to work efficiently. The study compared the performances between different forecasting models for the prediction of a time series formed by a group of dairy products. As performance measures, metric analysis of the various error-measuring techniques is done to select the least error-producing model for such products and hence to meet market demand. Authors compared the MA, regression, multiple regression, and Holt–Winters models based on MAPE, MAD, MSE, and RMSE applied for the demand forecasting of a time series formed by a group of dairy products at a milk processing unit located at northern India. The following conclusions have been drawn based on the outcomes of this study:

1. Multiple regression is being used globally for forecasting purpose, and the current study concludes that this method is suitable for the forecasting of dairy products such as milk, paneer, kheer, ice cream, table butter, and ghee because it produces least error; i.e., MAPE value was seen least while using this method.
2. Holt–Winters is also a suitable technique for the forecasting of dairy products. Based on the results obtained in the current study, we conclude that this method is useful for the forecasting of curd, plain lassi, and milk on the basis of daily demand since it also produces the least value of MAPE when compared with all other methods.
3. Multiple regression 2 is used in the current study for the data of the last three years to find out intercepts. This method may be useful for the forecasting of Pio bottle on the basis of error measures obtained.
4. Moving average is also a forecasting technique being used in various areas, but this technique is not suitable for the forecasting of dairy products because of high percentage error obtained.

Thus, it can be recommended that these methods of forecasting are very useful for milk processing industry and other short-lifecycle products. Further, this methodology may be applied to multiple case studies of food processing industry, though some other techniques like linear programming can also be attempted for the forecasting of such products.

6.5.1 Managerial/Research Implications

This is a unique study in itself which attempts to capture the dynamics of milk processing sector and to incorporate all relevant constraints that would significantly affect the demand and supply system. The study gives more visibility to forecasting theory by addressing the problem of dairy industry. Demand forecasting aided for attaining the high order fill rate, controlled inventory, and more process flexibility in the case industry. The forecasting model developed in this study has been discussed with the dairy industry managers and experts from academics. This study will help practitioners, regulators, and dairy industry professionals to focus their efforts in handling the demand fluctuations effectively, distribution channel management, and supply chain coordination by means of eradicating the uncertainties through demand forecasting. Further, this study should be of interest to researchers working in the area of operations management, marketing management, production planning and control, food supply chain management, etc.

Acknowledgements The authors would like to thank all the key resource persons from dairy industry. Further, the authors would like to express their sincere gratitude for the remarks and recommendations made by anonymous reviewers and editor which radically improved the quality of this work.

Appendix

Forecasting Models and the Calculations for MAD, MSE, MAPE, and RMSE

1. **Moving Average (MA)**

A. MA5: Moving average of 5 months (MA5) has been considered here, as derived below:

The values in Table 6.11 have been calculated for 'One period,' and rest can be calculated by using formula for MAPE, MAD, MSE, and RMSE.

(a) Forecast Demand, i.e., C11 = AVERAGE(B6:B10)
(b) Error = B11 − C11
(c) abs error = ABS(D11)
(d) Square Error = E11 * E11
(e) % Error = E11/B11 * 100

Table 6.11 MA5

	A	B	C	D	E	F	G	H	I	J	K
1	milk										
2											
3		moving average 5									
4											
5	t	y	Forecast	Error	abs error	sq. error	%error	MAPE	MAD	MSE	RMSE
6	1	3401307									
7	2	3562740									
8	3	3295195									
9	4	3506592									
10	5	3773399									
11	6	3577610	3507846.6	69763.4	69763.4	4866931980	1.95	3.820019	136844.9	26637481767	163209.9
12	7	3705007	3543107.2	161899.8	161899.8	26211545240	4.369757				
13	8	3424827	3571560.6	-146734	146733.6	21530749369	4.284409				
14	9	3424705	3597487	-172782	172782	29853619524	5.045164				
15	10	3446159	3581109.6	-134951	134950.6	18211664440	3.915971				
16	11	3304422	3515661.6	-211240	211239.6	44622168608	6.392634				
17	12	3454407	3461024	-6617	6617	43784689	0.191552				
18	13	3498093	3410904	87189	87189	7601921721	2.492472				
19	14	3664946	3425557.2	239388.8	239388.8	57306997565	6.531851				
20	15	3552001	3473605.4	78395.6	78395.6	6145870099	2.207083				
21	16	3610279	3494773.8	115505.2	115505.2	13341451227	3.199343				
22	17	3738022	3555945.2	182076.8	182076.8	33151961098	4.87094				
23	18	3674154	3612668.2	61485.8	61485.8	3780503602	1.673468				

(f) MAPE = AVERAGE(G11:G53)
(g) MAD = AVERAGE(E11:E53)
(h) MSE = AVERAGE(F11:F53)
(i) RMSE = SQRT(J11)

B. MA6: Here, the moving average of 6 months has been derived, as explained below:

The values in Table 6.12 have been calculated for 'One period,' and rest can be calculated by using formula for MAPE, MAD, MSE, and RMSE.

(a) Forecast Demand, i.e., C66 = AVERAGE(B60:B65)
(b) Error = B66 − C66
(c) abs error = ABS(D66)
(d) Square Error = E66 * E66
(e) % Error = E66/B66 * 100
(f) MAPE = AVERAGE(G66:G107)
(g) MAD = AVERAGE(E66:E107)
(h) MSE = AVERAGE(F66:F107)
(i) RMSE = SQRT(J66)

Table 6.12 MA6

	A	B	C	D	E	F	G	H	I	J	K
57		moving average 6									
58											
59	t	y	Forecast	Error	abs error	sq. error	%error	MAPE	MAD	MSE	RMSE
60	1	3401307									
61	2	3562740									
62	3	3295195									
63	4	3506592									
64	5	3773399									
65	6	3577610									
66	7	3705007	3519473.83	185533.2	185533.2	34422555933	5.007633	4.012007	143867.3	30353610176	174222.9
67	8	3424827	3570090.5	-145264	145263.5	21101484432	4.241484				
68	9	3424705	3547105	-122400	122400	14981760000	3.57403				
69	10	3446159	3568690	-122531	122531	15013845961	3.555582				
70	11	3304422	3558617.83	-254196	254195.8	64615521684	7.692596				
71	12	3454407	3480455	-26048	26048	678498304	0.754051				
72	13	3498093	3459921.17	38171.83	38171.83	1457088860	1.091218				
73	14	3664946	3425435.5	239510.5	239510.5	57365279610	6.535171				
74	15	3552001	3465455.33	86545.67	86545.67	7490152419	2.436533				
75	16	3610279	3486671.33	123607.7	123607.7	15278855259	3.42377				
76	17	3738022	3514024.67	223997.3	223997.3	50174805340	5.992403				
77	18	3674154	3586291.33	87862.67	87862.67	7719848194	2.391371				
78	19	3698504	3622915.83	75588.17	75588.17	5713570940	2.04375				
79	20	3479205	3656317.67	-177113	177112.7	31368896694	5.090607				

C. MA7: Here, the moving average of 7 months has been derived, as explained below:

The values in Table 6.13 have been calculated for 'One period,' and rest can be calculated by using formula for MAPE, MAD, MSE, and RMSE.

(a) Forecast Demand, i.e., C121 = AVERAGE(B114:B120)

(b) Error = B121 − C121

(c) abs error = ABS(D121)

(d) Square Error = E121 * E121

(e) % Error = E121/B121 * 100

(f) MAPE = AVERAGE(G121:G161)

(g) MAD = AVERAGE(E121:E161)

(h) MSE = AVERAGE(F121:F161)

(i) RMSE = SQRT(J121)

Table 6.13 MA7

	A	B	C	D	E	F	G	H	I	J	K
111	moving average 7										
112											
113	t	y	Forecast	Error	abs error	sq. error	%error	MAPE	MAD	MSE	RMSE
114	1	3401307									
115	2	3562740									
116	3	3295195									
117	4	3506592									
118	5	3773399									
119	6	3577610									
120	7	3705007									
121	8	3424827	3545978.57	-121152	121151.6	14677703260	3.537451	3.87788	139026.7	28730615087	169501.1
122	9	3424705	3549338.57	-124634	124633.6	15533527127	3.63925				
123	10	3446159	3529619.29	-83460.3	83460.29	6965619292	2.421835				
124	11	3304422	3551185.57	-246764	246763.6	60892260184	7.467677				
125	12	3454407	3522304.14	-67897.1	67897.14	4610022008	1.965522				
126	13	3498093	3476733.86	21359.14	21359.14	456212983.6	0.610594				
127	14	3664946	3465374.29	199571.7	199571.7	39828869143	5.44542				
128	15	3552001	3459651.29	92349.71	92349.71	8528469729	2.599935				
129	16	3610279	3477819	132460	132460	17545651600	3.668969				
130	17	3738022	3504329.57	233692.4	233692.4	54612151172	6.251767				
131	18	3674154	3546024.29	128129.7	128129.7	16417223683	3.487326				
132	19	3698504	3598843.14	99660.86	99660.86	9932286446	2.694626				
133	20	3479205	3633714.14	-154509	154509.1	23873075226	4.440932				

D. MA8: Here, the moving average of 8 months has been derived, as explained below:

The values in Table 6.14 have been calculated for 'One period,' and rest can be calculated by using formula for MAPE, MAD, MSE, and RMSE.

(a) Forecast Demand, i.e., C176 = AVERAGE(B168:B175)
(b) Error = B176 − C176
(c) abs error = ABS(D176)
(d) Square Error = E176 * E176
(e) % Error = E176/B176 * 100
(f) MAPE = AVERAGE(G176:G215)
(g) MAD = AVERAGE(E176:E215)
(h) MSE = AVERAGE(F176:F215)
(i) RMSE = SQRT(J176)

Table 6.14 MA8

	A	B	C	D	E	F	G	H	I	J	K
165			Moving average8								
166											
167	t	y	Forecast	Error	abs error	sq. error	%error	MAPE	MAD	MSE	RMSE
168	1	3401307									
169	2	3562740									
170	3	3295195									
171	4	3506592									
172	5	3773399									
173	6	3577610									
174	7	3705007									
175	8	3424827									
176	9	3424705	3530834.63	-106130	106129.6	11263497303	3.098942	3.839402	138103.3	28628503966	169199.6
177	10	3446159	3533759.38	-87600.4	87600.38	7673825700	2.541971				
178	11	3304422	3519186.75	-214765	214764.8	46123897843	6.499314				
179	12	3454407	3520340.13	-65933.1	65933.13	4347176972	1.908667				
180	13	3498093	3513817	-15724	15724	247244176	0.449502				
181	14	3664946	3479403.75	185542.3	185542.3	34425926535	5.062619				
182	15	3552001	3490320.75	61680.25	61680.25	3804453240	1.736493				
183	16	3610279	3471195	139084	139084	19344359056	3.852445				
184	17	3738022	3494376.5	243645.5	243645.5	59363129670	6.518033				
185	18	3674154	3533541.13	140612.9	140612.9	19771980616	3.827082				
186	19	3698504	3562040.5	136463.5	136463.5	18622286832	3.689695				

E. MA9: Here, the moving average of 9 months has been derived, as explained below:

The values in Table 6.15 have been calculated for one period, and rest can be calculated by using formula for MAPE, MAD, MSE, and RMSE.

(a) Forecast Demand, i.e., C230 = AVERAGE(B221:B229)
(b) Error = B230 − C230
(c) abs error = ABS(D230)
(d) Square Error = E230 * E230
(e) % Error = E230/B230 * 100
(f) MAPE = AVERAGE(G230:G268)
(g) MAD = AVERAGE(E230:E268)
(h) MSE = AVERAGE(F230:F268)
(i) RMSE = SQRT(J230)

Table 6.15 MA9

	A	B	C	D	E	F	G	H	I	J	K
218			Moving average9								
219											
220	t	y	Forecast	Error	abs error	sq. error	%error	MAPE	MAD	MSE	RMSE
221	1	3401307									
222	2	3562740									
223	3	3295195									
224	4	3506592									
225	5	3773399									
226	6	3577610									
227	7	3705007									
228	8	3424827									
229	9	3424705									
230	10	3446159	3519042.44	-72883.4	72883.44	5311996474	2.114918	3.648302	131416.6	27807111090	166754.6
231	11	3304422	3524026	-219604	219604	48225916816	6.645761				
232	12	3454407	3495324	-40917	40917	1674200889	1.184487				
233	13	3498093	3513014.22	-14921.2	14921.22	222642872.6	0.426553				
234	14	3664946	3512069.89	152876.1	152876.1	23371105348	4.171306				
235	15	3552001	3500019.56	51981.44	51981.44	2702070567	1.463441				
236	16	3610279	3497174.11	113104.9	113104.9	12792715891	3.132857				
237	17	3738022	3486648.78	251373.2	251373.2	63188496850	6.724766				
238	18	3674154	3521448.22	152705.8	152705.8	23319054567	4.156216				
239	19	3698504	3549164.78	149339.2	149339.2	22302203294	4.037828				
240	20	3479205	3577203.11	-97998.1	97998.11	9603629781	2.816681				

2. Holt and Winters

A. Monthly Demand

The values in Table 6.16 have been calculated for 'One period,' and rest can be calculated by using formula for MAPE, MAD, MSE, and RMSE.

(a) Forecast Demand

- For first period

  Level Factor (u), D15 = C15/F15
  Trend Factor (v), E15 = 0
  Seasonal Factor (s), F4 = C4/AVERAGE(C$4:C$15)

- For the other periods using the following formula

  (u), D16 = M$14 * C16/F4 + (1 − M$14) * (D15 + E15)
  (v), E16 = O$14 * (D16 − D15) + (1 − O$14) * E15
  (s), F16 = Q$14*C16/D16 + (1 − Q$14) * F4
  So, the Forecast, G16 = (D15 + E15) * F4
  (b) Error = C16 − G16
  (c) abs error = ABS(H16)
  (d) Square Error = I16 * I16
  (e) % Error = I16/C16 * 100
  (f) MAPE = AVERAGE(K16:K51)

Table 6.16 Monthly demand

	A	B	C	D	E	F	G	H	I	J	K	L	M	N	O	P	Q
1	Milk																
2																	
3		t	y	u	v	s	Forecast	Error	Abs. Error	Square Error	% error	MAD	MSE	MAPE	RMSE		
4		1	3401307			0.974671						55706.5	5411244645	1.552632	73561.16		
5		2	3562740			1.020931											
6		3	3295195			0.944264											
7		4	3506592			1.004841											
8		5	3773399			1.081297											
9		6	3577610			1.025192											
10		7	3705007			1.061699											
11		8	3424827			0.981411											
12		9	3424705			0.981376											
13		10	3446159			0.987524											
14		11	3304422			0.946908						alpha		0.04 beta		0.95 gama		0.2
15		12	3454407	3489698	0	0.989887											
16		13	3498093	3493670	3773.446	0.97999	3401307	96786	96786	9367529796	2.766822						
17		14	3664946	3501138	7283.319	1.026102	3570648	94298.39	94298.386	8892185672	2.572982						
18		15	3552001	3518551	16906.47	0.957312	3312875	239126.2	239126.16	57181318118	6.732153						
19		16	3610279	3537754	19088.73	1.007973	3552573	57705.94	57705.945	3329976069	1.598379						
20		17	3738022	3552848	15293.95	1.075461	3846003	-107981	107981.08	11659913610	2.888722						
21		18	3674154	3568771	15891.57	1.026059	3658031	16122.99	16122.994	259950929.3	0.438822						
22		19	3698504	3580619	12050.13	1.055943	3805832	-107328	107327.71	11519236621	2.901922						
23		20	3479205	3590767	10242.7	0.978915	3525885	-46679.8	46679.817	2179005350	1.341681						

(g) MAD = AVERAGE(I16:I51)
(h) MSE = AVERAGE(J16:J51)
(i) RMSE = SQRT(M4)

B. Daily Demand

The values in Table 6.17 have been calculated for 'One period,' and rest can be calculated by using formula for MAPE, MAD, MSE, and RMSE.

(a) Forecast Demand

- For first period

Level Factor (u), D9 = C9/F9
Trend Factor (v), E9 = 0
Seasonal Factor (s), F3 = C3/AVERAGE(C$3:C$9)

- For the other periods using the following formula

(u), D10 = M$16 * C10/F3 + (1 − M$16) * (D9 + E9)
(v), E10 = O$16 * (D10 − D9) + (1 − O$16) * E9
(s), F10 = Q$16*C10/D10 + (1 − Q$16) * F3
So, the Forecast, G10 = (D9 + E9) * F3
(b) Error = C10 − G10
(c) abs error = ABS(H10)
(d) Square Error = I10 * I10
(e) % Error = I10/C10 * 100
(f) MAPE = AVERAGE(K10:K1463)
(g) MAD = AVERAGE(I10:I1463)

Table 6.17 Daily demand

	A	B	C	D	E	F	G	H	I	J	K	L	M	N	O	P	Q
1	Milk																
2		t	y	u	v	s	Forecast	Error	Abs. Error	Square Err	% error	MAD	MSE	MAPE	RMSE		
3	Monday	1	109699			0.995049							3134.495	52411611	2.994861	7239.586	
4	Tuesday	2	110792			1.004963											
5	Wednesd:	3	110876			1.005725											
6	Thursday	4	110903			1.00597											
7	Friday	5	110725			1.004355											
8	Saturday	6	111518			1.011548											
9	Sunday	7	107201	110244.9	0	0.97239											
10		8	110704	110446.6	0	0.995121	109699	1005	1005	1010025	0.907826						
11		9	114084	111061.6	0	1.005185	110995	3088.997	3088.997	9541904	2.707652						
12		10	114203	111559.9	0	1.005905	111697.4	2505.576	2505.576	6277909	2.193966						
13		11	116107	112331.5	0	1.006246	112225.9	3881.138	3881.138	15063235	3.342726						
14		12	118877	113537.5	0	1.004782	112820.7	6056.28	6056.28	36678524	5.094577						
15		13	124197	115385.8	0	1.012196	114848.7	9348.34	9348.34	87391453	7.527025						
16		14	113675	115689.2	0	0.972492	112200	1474.983	1474.983	2175574	1.297544	alpha		0.2	beta	0	gama	0.01
17		15	116092	115883.6	0	0.995188	115124.8	967.2034	967.2034	935482.5	0.833135						
18		16	116314	115849.7	0	1.005174	116484.5	-170.49	170.4903	29066.95	0.146578						
19		17	117840	116109.4	0	1.005995	116533.7	1306.299	1306.299	1706418	1.108536						
20		18	119683	116675.5	0	1.006442	116834.6	2848.372	2848.372	8113222	2.37993						
21		19	116590	116547.4	0	1.004738	117233.5	-643.459	643.4592	414039.8	0.551899						
22		20	108522	114680.8	0	1.011537	117968.9	-9446.91	9446.907	89244057	8.705062						

Table 6.18 Regression analysis

y	t	prediction	error	abs error	square error	% error										MAD	MSE	MAPE	RMSE
3401307	1	3518528.486	-117221	117221.5	13740876673	3.446366													
3562740	2	3521136.144	41603.86	41603.86	1730880874	1.167749		SUMMARY OUTPUT								117644.3	21625623286	3.28351915	147056.5
3295195	3	3523743.801	-228549	228548.8	52234554663	6.93582													
3506592	4	3526351.459	-19759.5	19759.46	390436238.3	0.563495		Regression Statistics											
3773399	5	3528959.117	244439.9	244439.9	59750856187	6.477976		Multiple F	0.238561										
3577610	6	3531566.775	46043.22	46043.22	2119978531	1.286983		R Square	0.056911										
3705007	7	3534174.433	170832.6	170832.6	29183765817	4.610857		Adjusted I	0.036409										
3424827	8	3536782.091	-111955	111955.1	12533942481	3.268927		Standard I	150219.4										
3424705	9	3539389.749	-114685	114684.7	13152591729	3.348748		Observati	48										
3446159	10	3541997.407	-95838.4	95838.41	9185000315	2.781021													
3304422	11	3544605.065	-240183	240183.1	57687904846	7.268535		ANOVA											
3454407	12	3547212.723	-92805.7	92805.72	8612902268	2.686589			df	SS	MS	F	gnificance F						
3498093	13	3549820.381	-51727.4	51727.38	2675721968	1.478731		Regressio	1	6.26E+10	6.26E+10	2.7758957	0.10249						
3664946	14	3552428.039	112518	112518	12660291503	3.070112		Residual	46	1.04E+12	2.26E+10								
3552001	15	3555035.697	-3034.7	3034.697	9209386.916	0.085436		Total	47	1.1E+12									
3610279	16	3557643.355	52635.64	52635.64	2770511109	1.457938													
3738022	17	3560251.013	177771	177771	31602523777	4.75575		Coefficient	andard Err	t Stat	P-value	Lower 95%	Upper 95%	Lower 95.0%	Upper 95.0%				
3674154	18	3562858.671	111295.3	111295.3	12386650237	3.029142		Intercept	3515921	44051.16	79.81449	5.565E-51	3427250	3604591	3427250.444	3604591.211			
3698504	19	3565466.329	133037.7	133037.7	17699021888	3.597067		t	2607.658	1565.125	1.666102	0.1024896	-542.775	5758.091	-542.775442	5758.091389			
3479205	20	3568073.987	-88969	88868.99	7897696857	2.55429													
3604801	21	3570681.645	34119.35	34119.35	1164130385	0.946498													

  (h) MSE = AVERAGE(J10:J1463)
  (i) RMSE = SQRT(M3)

3. Regression Analysis

The values in Table 6.18 have been calculated for 'One period,' and rest can be calculated by using formula for MAPE, MAD, MSE, and RMSE.
  (a) Forecast Demand, F3 = M$20 + D3 * M$21
  (b) Error = C3 − F3
  (c) abs error = ABS(G3)
  (d) Square Error = H3 * H3
  (e) % Error = H3/C3 * 100
  (f) MAPE = AVERAGE(J3:J50)
  (g) MAD = AVERAGE(H3:H50)
  (h) MSE = AVERAGE(I3:I50)
  (i) RMSE = SQRT(S4)

4. Multiple Regression

A. Monthly Demand

The values in Tables 6.19 and 6.20 have been calculated for 'One period,' and rest can be calculated by using formula for MAPE, MAD, MSE, and RMSE.
  (a) Forecast Demand, P57

Table 6.19 Monthly demand

	C	D	E	F	G	H	I	J	K	L	M	N	O	P	Q	R	S	T
54	multiple regression																	
55																		
56	y	t	M1	M2	M3	M4	M5	M6	M7	M8	M9	M10	M11	prediction	error	abs error	% error	square error
57	3401307	1	1	0	0	0	0	0	0	0	0	0	0	3414673.4	-13366.4	13366.44	0.392979633	178661818.5
58	3562740	2	0	1	0	0	0	0	0	0	0	0	0	3571989.1	-9249.07	9249.069	0.259605493	85545272.74
59	3295195	3	0	0	1	0	0	0	0	0	0	0	0	3403208.7	-108014	108013.7	3.277915078	11666958038
60	3506592	4	0	0	0	1	0	0	0	0	0	0	0	3546265.4	-39673.4	39673.44	1.131396061	1573982139
61	3773399	5	0	0	0	0	1	0	0	0	0	0	0	3760314.2	13084.81	13084.81	0.346764449	171212154.6
62	3577610	6	0	0	0	0	0	1	0	0	0	0	0	3613416.2	-35806.2	35806.19	1.000841169	1282083511
63	3705007	7	0	0	0	0	0	0	1	0	0	0	0	3748727.9	-43720.9	43720.94	1.180050233	1911520922
64	3424827	8	0	0	0	0	0	0	0	1	0	0	0	3434717.7	-9890.69	9890.694	0.288793967	97825822.86
65	3424705	9	0	0	0	0	0	0	0	0	1	0	0	3468790.7	-44085.7	44085.69	1.287284416	1943548393
66	3446159	10	0	0	0	0	0	0	0	0	0	1	0	3507407.4	-61248.4	61248.44	1.777295933	3751371862
67	3304422	11	0	0	0	0	0	0	0	0	0	0	1	3276619.9	27802.06	27802.06	0.841359132	772954331.7
68	3454407	12	0	0	0	0	0	0	0	0	0	0	0	3500149.9	-45742.9	45742.94	1.324190918	2092416903
69	3498093	13	1	0	0	0	0	0	0	0	0	0	0	3454196.8	43896.19	43896.19	1.254860446	1926875094
70	3664946	14	0	1	0	0	0	0	0	0	0	0	0	3611512.4	53433.56	53433.56	1.457963103	2855145379
71	3552001	15	0	0	1	0	0	0	0	0	0	0	0	3442732.1	109268.9	109268.9	3.076264208	11939700247
72	3610279	16	0	0	0	1	0	0	0	0	0	0	0	3585788.8	24490.19	24490.19	0.678346062	599769181.7
73	3738022	17	0	0	0	0	1	0	0	0	0	0	0	3799837.6	-61815.6	61815.56	1.653697185	3821164025
74	3674154	18	0	0	0	0	0	1	0	0	0	0	0	3652939.6	21214.44	21214.44	0.577396468	450052270
75	3698504	19	0	0	0	0	0	0	1	0	0	0	0	3788251.3	-89747.3	89747.31	2.426584224	8054580475
76	3479205	20	0	0	0	0	0	0	0	1	0	0	0	3474241.1	4963.935	4963.935	0.142674416	24640654.82

Table 6.20 ANOVA

	A	B	C	D	E	F	G	H	I	J
118										
119		ANOVA								
120			df	SS	MS	F	gnificance F			
121		Regressio	12	9.6E+11	8E+10	19.92460831	3.57E-12			
122		Residual	35	1.41E+11	4.02E+09					
123		Total	47	1.1E+12						
124										
125			Coefficients	andard Err	t Stat	P-value	Lower 95%	Upper 95%	Lower 95.0%	pper 95.0%
126		Intercept	3460627	37712.18	91.76415	2.65318E-43	3384067	3537186	3384066.767	3537186
127		t	3293.614	681.7435	4.831164	2.66887E-05	1909.601	4677.627	1909.601431	4677.627
128		M1	-49246.7	45431.95	-1.08397	0.285790161	-141479	42985.02	-141478.503	42985.02
129		M2	104775.3	45324.41	2.311674	0.026802606	12761.83	196788.7	12761.83257	196788.7
130		M3	-67298.7	45226.88	-1.48802	0.145699596	-159114	24516.73	-159114.177	24516.73
131		M4	72464.41	45139.45	1.605346	0.117404221	-19173.5	164102.4	-19173.5394	164102.4
132		M5	283219.5	45062.16	6.285086	3.25784E-07	191738.5	374700.6	191738.5018	374700.6
133		M6	133027.9	44995.07	2.9565	0.005540018	41683.09	224372.8	41683.0903	224372.8
134		M7	265046.1	44938.22	5.89801	1.05206E-06	173816.6	356275.5	173816.6334	356275.5
135		M8	-52257.8	44891.66	-1.16409	0.252262827	-143393	38877.11	-143392.698	38877.11
136		M9	-21478.4	44855.4	-0.47884	0.635032945	-112540	69582.9	-112539.719	69582.9
137		M10	13844.73	44829.49	0.308831	0.759281117	-77164	104853.4	-77163.9799	104853.4
138		M11	-220236	44813.94	-4.91446	2.07673E-05	-311214	-129259	-311213.518	-129259

P57 = C$126 + C$127 * D57 + C$128 * E57 + C$129*F57 + C
$130 * G57 + C$131 * H57 + C$132 * I57 + C$133 * J57 + C$134 * K57 + C
$135 * L57 + C$136 * M57 + C$137 * N57 + C$138 * O57

(b) Error = C57 − P57

(c) abs error = ABS(Q57)

(d) Square Error = R57 * R57

(e) % Error = R57/C57 * 100

(f) MAPE = AVERAGE(S57:S104)
(g) MAD = AVERAGE(R57:R104)
(h) MSE = AVERAGE(T57:T104)
(i) RMSE = SQRT(W57)

B. Daily Demand

The values in Tables 6.21 and 6.22 have been calculated for 'One period,' and rest can be calculated by using formula for MAPE, MAD, MSE, and RMSE.

(a) Forecast Demand, J4

J4 = Q\$22 + C4 * Q\$23 + D4 * Q\$24 + E4 * Q\$25 + F4 * Q\$26 + G4 * Q\$27 + H4 * Q\$28 + I4 * Q\$29

(b) Error = B4 − J4
(c) abs error = ABS(K4)
(d) Square Error = L4 * L4
(e) % Error = L4/B4 * 100
(f) MAPE = AVERAGE(N4:N1464)

Table 6.21 Daily demand

	A	B	C	D	E	F	G	H	I	J	K	L	M	N	O	P	Q	R	S	T
1																				
2											predictior error		abs error	square error	% error			mad	mse	mape	RMSE
3	y	t	W1	W2	W3	W4	W5	W6									4484.643	59452295	4.070971	7710.531
4	109699	1	1	0	0	0	0	0	0	116263.9	-6564.91	6564.909	43098025.58	5.984474						
5	110792	2	0	1	0	0	0	0	0	116322.9	-5530.91	5530.911	30590976.95	4.992157						
6	110876	3	0	0	1	0	0	0	0	116212.3	-5336.34	5336.339	28476516.82	4.812889		SUMMARY OUTPUT				
7	110903	4	0	0	0	1	0	0	0	114915.1	-4012.15	4012.145	16097311.45	3.617707						
8	110725	5	0	0	0	0	1	0	0	115500.8	-4775.78	4775.782	22808092.33	4.313192		Regression Statistics				
9	111518	6	0	0	0	0	0	1	1	115855.4	-4337.38	4337.381	18812871.96	3.8894		Multiple F 0.220215				
10	107201	7	0	0	0	0	0	0	0	112684.3	-5483.26	5483.261	30066146.58	5.114934		R Square 0.048494				
11	110704	8	1	0	0	0	0	0	0	116284.8	-5580.81	5580.809	31145430.16	5.041199		Adjusted I 0.04391				
12	114084	9	0	1	0	0	0	0	0	116343.8	-2259.81	2259.811	5106747.961	1.980831		Standard I 7731.729				
13	114203	10	0	0	1	0	0	0	0	116233.2	-2030.24	2030.24	4121873.311	1.777746		Observati 1461				
14	116107	11	0	0	0	1	0	0	114936	1170.954	1170.954	1371133.416	1.008513							

Table 6.22 ANOVA

	O	P	Q	R	S	T	U	V	W	X
14										
15		ANOVA								
16			df	SS	MS	F	gnificance F			
17		Regressio	7	4.43E+09	6.32E+08	10.57909	4.94E-13			
18		Residual	1453	8.69E+10	59779630					
19		Total	1460	9.13E+10						
20										
21			Coefficients	andard Err	t Stat	P-value	Lower 95%	Upper 95%	ower 95.0%	pper 95.0%
22		Intercept	112663.4	640.6951	175.8455	0	111406.6	113920.1	111406.6	113920.1
23		t	2.985778	0.479616	6.225348	6.27E-10	2.044964	3.926592	2.044964	3.926592
24		W1	3597.563	757.2519	4.750814	2.23E-06	2112.139	5082.987	2112.139	5082.987
25		W2	3653.579	757.2513	4.824791	1.55E-06	2168.157	5139.002	2168.157	5139.002
26		W3	3540.022	757.251	4.674833	3.21E-06	2054.6	5025.444	2054.6	5025.444
27		W4	2239.842	757.251	2.95786	0.003148	754.4202	3725.264	754.4202	3725.264
28		W5	2822.493	757.2513	3.727287	0.000201	1337.07	4307.915	1337.07	4307.915
29		W6	3174.106	758.1585	4.186599	3E-05	1686.904	4661.308	1686.904	4661.308

(g) MAD = AVERAGE(L4:L1464)
(h) MSE = AVERAGE(M4:M1464)
(i) RMSE = SQRT(R4)

References

Amorim, P., Alem, D., & Lobo, B. A. (2013). Risk management in production planning of perishable goods. *Industrial and Engineering Chemistry Research, 52*(49), 17538–17553.

Assis, M. V. O. D, Carvalho, L. F, Rodrigues, J. J., & Proença, M. L. (2013). Holt-Winters statistical forecasting and aco metaheuristic for traffic characterization. In *Communication QoS, Reliability and Modeling Symposium,* 978-1-4673-3122-7/13.

Bhardwaj, A., Mor, R. S., Singh, S., & Dev, M. (2016). An investigation into the dynamics of supply chain practices in dairy industry: A pilot study. In *Proceedings of the 2016 International Conference on Industrial Engineering and Operations Management Detroit, Michigan, USA, September 23–25, 2016* (pp. 1360–1365).

Dhahri, I., & Chabchoub, H. (2007). Nonlinear goal programming models quantifying the bullwhip effect in supply chain based on ARIMA parameters. *European Journal of Operational Research, 177*(3), 1800–1810.

García, J. A., Garrido, G. P., & Prado, J. C. (2014). Packaging as source of efficient and sustainable advantages in supply chain management: an analysis of milk cartons. *International Journal of Production Management, 2*(1), 15–22.

Ghosh, S. (2008). Univariate time-series forecasting of monthly peak demand of electricity in northern India. *International Journal of Indian Culture and Business Management, 1*(4), 466–474.

Gupta, S. (2015). ECMS based hybrid algorithm for energy management in parallel hybrid electric vehicles. *HCTL Open International Journal of Technology Innovations and Research, 14,* 1–12.

Harsoor, A. S., & Patil, A. (2015). Forecast of sales of Walmart store using big data applications. *International Journal of Research in Engineering and Technology, 4*(6), 51–59.

Hassan, M., Eldin, A. B., & Ghazali, A. E. (2015). A decision support system for subjective forecasting of new product sales. *International Journal of Computer Applications, 126*(2), 25–30.

Jraisat, L., Gotsi, M., & Bourlakis, M. (2013). Drivers of information sharing and export performance in the Jordanian agri-food export supply chain. *International Marketing Review, 30*(4), 323–356.

Kaloxylos, A., Wolfert, J., Verwaart, T., Terol, C. M., Brewster, C., Robbemond, R., et al. (2013). The use of future internet technologies in agri. & food sectors: integrating the supply chain. *Procedia Technology, 8,* 51–60.

Leat, P., & Giha, C. R. (2008). Building collaborative agri-food supply chains. *British Food Journal, 102*(4), 311–395.

Mishra, B. K., Bharadi, V. A., Nemade, B., Vhatkar, S., & Dias, J. (2016). Oral-care goods sales forecasting using artificial neural network model. *Procedia Computer Science, 79,* 238–243.

Mor, R. S., Bhardwaj, A., & Singh, S. (2018a). Benchmarking the interactions among barriers in dairy supply chain: An ISM approach. *International Journal for Quality Research, 12*(2), 385–404.

Mor, R. S., Bhardwaj, A., & Singh, S. (2018b). A structured-literature-review of the supply chain practices in dairy industry. *Journal of Operations and Supply Chain Management, 11*(1), 14–25.

Mor, R. S., Bhardwaj, A., & Singh, S. (2018c). Benchmarking the interactions among performance indicators in dairy supply chain: An ISM approach. *Benchmarking: An International Journal* (in press).

Mor, R. S., Bhardwaj, A., & Singh, S. (2018d). A structured literature review of the supply chain practices in food processing industry. In *Proceedings of the 2018 International Conference on Industrial Engineering and Operations Management, Bandung, Indonesia, March 6–8, 2018* (pp. 588–599).

Mor, R.S., Nagar, J., & Bhardwaj, A. (2018e). A comparative study of forecasting methods for sporadic demand in an auto service station. *International Journal of Business Forecasting & Marketing Intelligence* (in press).

Mor, R. S., Singh, S., Bharadwaj, A., & Singh, L. P. (2015). Technological implications of supply chain practices in agri-food sector: A review. *International Journal of Supply and Operations Management, 2*(2), 720–747.

Mor, R. S., Singh, S., Bhardwaj, A., & Bharti, S. (2017). Exploring the causes of low-productivity in dairy industry using AHP. *Jurnal Teknik Industri, 19*(2), 83–92.

Patushi, S., & Kume, V. (2014). The development of clusters as a way to increase competitiveness of Businesses (case of milk processing industry in Tirana). *European Scientific Journal, 10*(13), 98–116.

Ping, L. (2016). Based on the regression analysis model throughput forecast of port group in the upper Yangtze River. *School of Economics and Management.* ISSN: 978-1-5090-1102-5/16.

Sarno, R., & Herdiyanti, A. (2010). A service portfolio for an enterprise resource planning. *International Journal of Computer Science and Network Security, 10*(3), 144–156.

Spicka, J. (2013). The competitive environment in the dairy industry and its impact on the food industry. *Agris On-line Papers in Economics and Informatics, 5*(2), 89–102.

Sugiarto, V. C., Sarno, R., & Sunaryono, D. (2016). Sales forecasting using holt-winters in enterprise resource planning at sales and distribution module. In *International Conference on Information, Communication Technology and System.* IEEE (pp. 8–13).

Taylor, J. W. (2011). Multi-item sales forecasting with total and split exponential smoothing. *Journal of the Operational Research Society, 62,* 555–563.

Tratar, L. F., & Strmčnik, E. (2016). The comparison of Holt-Winters method and multiple regression method: A case study. *Energy, 109,* 266–276.

Vaida P. (2008). Selection of market demand forecast methods: criteria and application. *Engineering Economics, Economics of Engineering Decisions, 3*(58).

Veiga, C. P., Veiga, C. R., Catapan, A., Tortato, U., & Silva, W. V. (2014). Demand forecasting in food retail: A comparison between the Holt-Winters and ARIMA models. *WSEAS Transactions on Business and Economics, 11,* 608–614.

Weber, S. A., Salamon, P., & Hansen, H. (2014). Policy impacts in the dairy supply chain: The case of German whole milk powder. In *Proceedings in System dynamics and Innovation in Food networks* (pp. 439–447).

Yuan, X. M., & Cai, T. X. (2008). New forecasting algorithms for intermittente demands. *SIMTech. Technical Reports, 9*(4), 228–232.

Zhou, Q., Han, R., & Li, T. (2015). A two-step dynamic inventory forecasting model for large manufacturing. In *International Conference on Machine Learning and Applications* (Vol. 2, No. 4, pp. 749–753).

Chapter 7
Customer Experience and Its Marketing Outcomes in Financial Services: A Multivariate Approach

Swati Raina, Hardeep Chahal, Phillip Klaus and Kamani Dutta

Abstract The purpose of this paper is twofold: first, to validate the customer experience quality (EXQ) scale in Indian sector, across financial products in Indian settings. And second, to assess EXQ impact on the marketing outcomes, that is customer satisfaction, word-of-mouth, loyalty intentions and service value. The respondents comprised of customers of Jammu City, India, who have experienced one of the three services, that is *Banking or Insurance or Investment* from the public and private sectors. The customer experience quality scale is assessed through validity and reliability analysis assuring the validation of EXQ scale in Indian settings. It is validated as four-dimensional scale, that is peace of mind, moment of truth, outcome focus and product experience. Also, the findings suggest that all the four individual dimensions have positive and significant impact on the marketing outcomes. First, the study is based on three financial services such as banking, investment and insurance only, and for further research, it is suggested to adopt other services comprehensively to understand customer experience from their perspective. Secondly, the major limitation of the research is related to the presence of subjective responses of the customers with respect to customer experience constructs in the study. This study contributes to the extant marketing literature by

S. Raina
Lovely Professional University, Jalandhar - Delhi G.T. Road, Phagwara, Punjab, India
e-mail: rainaswati87@yahoo.in

H. Chahal (✉)
University of Jammu, Jammu, India
e-mail: chahalhardeep@rediffmail.com

P. Klaus
School of Management Centre for Advanced Research in Marketing,
Cranfield University, Bedfordshire, UK
e-mail: philipp.klaus@cranfield.ac.uk

K. Dutta
University of Jammu, Udhampur Campus, Jammu, India
e-mail: duttakamini11@gmail.com

© Springer Nature Singapore Pte Ltd. 2019
H. Chahal et al. (eds.), *Understanding the Role of Business Analytics*,
https://doi.org/10.1007/978-981-13-1334-9_7

119

validating the domain of consumer experience quality in financial service sector operating in emerging economies, that is India, and its impact on marketing outcomes.

Keywords Customer experience · Customer satisfaction · Moment of truth Peace of mind

7.1 Introduction

The term customer experience was first addressed in the literature by Pine and Gilmore (1999) and Carbone and Haeckel (1994). Since then, it has started receiving paramount attention from practitioners, academicians and consultants. And this is primarily because favourable experiences are considered as a base for companies to attract customers, differentiate themselves from competitors, achieve competitive advantage and make profit (Pareigis et al. 2011). Remarked that vast numbers of organizations are applying customer experience-based strategies to compete successfully in the market vis-a-vis to generate, strengthen and sustain customer loyalty. The significant attention on customer experience has resulted from two reasons. First, the distinction between experience and commodities has primarily shifted the focus of companies from service-based economy to an experience-based economy which has led to increasing interest among both academicians and practitioners on customer experience (Verhoef et al. 2009). This shift enhances the need of companies to deliver high levels of service quality to achieve outcomes—customer satisfaction, loyalty and positive word-of-mouth. Second, according to Knutson et al. (2006), though from different perspective, factors such as state-of-the-art technology, more sophisticated and demanding consumers, increasingly competitive business environment have made commodities to be similar to services. Even the recent contemporary school on service-dominant logic also established the dominance of services even among the products. Carbone and Haeckel (1994) suggested that whenever a customer purchases a service, he will have an experience that means every experience comes with the purchase of service. Further, in their paper on conceptualizing and measuring customer experience quality (EXQ) categorically remarked that "products and services might not be the most important offerings anymore; experience, which represents customers' personal sensations and fulfils customers' inner needs, is becoming a key element of a new economic stage".

Since each customer's experience is unique and individualized, the businesses must go beyond goods and services and start creating memorable experiences for each customer (Gilmore and Pine 2002). Thus, it is important for companies to make memories and create the stage for greater economic value rather than simply making goods and delivering services (Kim et al. 2011). However, observed that existing literature on the impact of customer experience on consumer behaviour is largely descriptive and generally explores the "what" rather than the "why". Also, firms are still typically measuring CE against service quality criteria, which has proved to be an insufficient approach (Klaus and Maklan 2011). Recently, Klaus and Maklan (2012) developed a measure of customer experience quality (EXQ), which needs to be validated across different service products. Hence, the purpose of this paper is twofold: first, to validate the customer experience quality (EXQ) scale of Klaus and Maklan (2012) developed and tested in UK settings, across financial products in Indian settings. And second, to assess EXQ impact on the marketing outcomes, that is customer satisfaction, word-of-mouth, loyalty intentions and service value. Adhikari and Bhattacharya (2015) give insight and direction into an understanding of customer experience from consumption of an experience product as well as interaction with sensory memorabilia. This remarked that this would help the marketer to categorize customer experience in a more meaningful way than considering it as overall marketing phenomena. Very recently, Rias et al. (2016) remarked that it is important for the marketer to learn and understand how to evaluate customer experience quality in determining the client future behaviour.

The study is a contribution to the extant literature on customer experience quality. It validates consumer experience quality in financial sector (i.e. banking, insurance and investment) in an emerging context, that is India. Further, it also compares the results of finance-based service experiences in Indian and UK contexts.

The paper is based on following structure. At the outset, the review of the literature on customer/service experience is discussed. In the next following section, a framework is presented for investigating key dimensions for creating favourable service or customer experiences. The next section discusses the methodology adopted for data collection, which is followed by data analysis. The final section summarizes the major findings, discusses the managerial and theoretical implications and concludes with suggestions for future research.

7.2 Review of Literature and Hypotheses Development

Customer experience in the literature is discussed in terms of impressions, feelings and intangible assets which results from the interaction between customer and service provider. It is defined as the user's interpretation of his total interactions with the brand, perceived value of the encounter/encounters between a customer and a product or a company or part of its organization which provokes his reaction (Biedenbach and Marell 2009; Verhoef et al. 2009; Frow and Payne 2007).

Most of the researchers consider customer experience from comprehensive perspective. According to customer experience focuses on direct experience of different individuals with varied frontline employees. In other words, it is a total experience including search, purchase, consumption and after-sale phases of experience. The scholars further remarked that effective personal interaction between a customer and an organization results in positive and stronger experience-based beliefs, which have positive impact on overall brand awareness. Similarly, Carbone and Haeckel (1994) defined "experience as a take away impression formed by people's encounters with products, services, and businesses—perception produced when humans consolidate sensory information" (p. 9). Haeckel et al. (2003) explained experience as the feelings that a customer gets after using the particular product, service, etc. They also mentioned various stages of delivery process which ultimately results in an experience.

Researchers namely Ding et al. (2011) and Nagasawa (2008) shared similar viewpoint on customer experience. They conceptualized customer experience as an important intangible asset which captures customer interaction within certain service systems, that occurs in response to some stimulation. On the flip side, Meyer and Schwager (2007) expressed customer experience in terms of internal and subjective responses, that results when a customer establishes direct and indirect contacts with a company. Direct contact is initiated by a customer and leads to experience during the course of purchase, use and service, whereas indirect contact involves unplanned encounters with a representative of a company's product, services or brands and takes the form of word-of-mouth or criticism. Customer experience is holistic in nature. The authors further expressed that customer experience involves responses namely cognitive, affective, emotional, social and physical. From significance perspective, Chung and Kwon (2009) stated that customer experience is important in understanding customer perceptions, attitudes and behaviour. According to customer experience comes from the interaction of customer with the firm, its staff, self-service technologies, service environment and service companies. For instance, Gilmore and Pine 2002 argued that it is important to realize that actual experiences are distinct from services. They stated that "when a person buys a service, he purchases a set of intangible activities carried out on his behalf. But when he buys an experience, he pays to spend time enjoying a series of memorable events that a company stages—as a theatrical play to engage him in a personal way" (p. 2). Similarly, Terblanche and Boshoff (2004) stated that "In-store shopping experiences include to all interactions and experiences that a customer goes through from entering to leaving the shop door" (p. 3). Helkkula (2011) expressed service experience to be a three-component framework based on phenomenological, process-based and outcome-based characteristics. He related service experience with hedonic expression. Similarly, Grace and Cass (2004) described service experience in terms of core service, employee service and servicescape. However, researchers like Frow and Payne (2007) considered customer experience as an outcome of co-creation process which is linked with a brand. They remarked that when a co-creation approach is adopted, customer engages himself in a dialogue and interaction with suppliers during product design, production, delivery and subsequent consumption. Rias et al. (2016), defined customer experience from emotional perspective. Suggest that emotions define the importance of an experience. Similarly, Rias et al. (2016) conceptualized customer experience quality as an overall customer journey through emotion towards product, services or even brand, which is at affective level.

Klaus and Maklan defined customer experience as the direct and indirect interactions (cognitive and emotional) of customer with the firm. This definition is highly consistent with conceptualizations offered by other researchers (e.g. Verhoef et al. 2009). Identified and evaluated critical success factors like servicescape, online functional elements, convenience and customer interaction for measuring customer experience in a service organization more effectively.

7.2.1   Customer Experience Dimensions

The customer experience is considered as a multidimensional construct in the literature by marketing researchers. However, the conceptualization has been considered from different perspectives. Its dimensionality was explained initially by Schmitt (2000, 1999) who discussed five types of experiences, which include sensory (sense), affective (feel), cognitive (think), physical experience (act) and social identity experience (relate). He expressed that sense marketing appeals to the senses with the objective of creating sensory experiences through sight, sound, touch, taste and smell. From functionality perspective, three clues namely functional, humanics and mechanics are considered by Haeckel et al. (2003) that lead to customer experience. These clues can be considered as three dimensions of customer experience. Functional clues are linked to the reliability of services. Berry et al. (2006) link functional clues with technical quality. They address that it is important for the service providers to manage functional clues to meet their expectations. The mechanics is referred to service encounters, service environment, servicescape, atmospherics, physical environment and mechanic clues. It involves place where service is assembled, service provider–customer interactions, combined with tangibles that facilitate performance or communication of the service. Mechanic clues are physical representation of services and include building, design, equipment, sound, smell or any other stimuli that communicate about the service without any word and action. Berry and Seltman (2007) mention that mechanic clues in certain situations have the ability to influence customers before humanics and functional clues and hence contribute to creating first-hand expression. Employee's behaviour and performance represent the third category of clues that is humanics. The humanic clues are concerned with action and appearance of employees and service providers such as tone of voice, body language, appearance, level of enthusiasm, etc. The customer–provider interactions are very significant and may contribute in high performance than their expectations and create an emotional involvement add service experience. Humanic clue behaviours include behaviour and performance of employee, perceived efforts of employee, credibility, competence. All the three clues play synergistic role in creating customer experience. Later, Payne et al. (2008) suggested that customer experience consists of three components: cognition, emotions and behaviour. These components are essential inputs to customer learning. They expressed that supplier can enhance the customer experience by supporting customer learning and developing processes that acknowledge cognitions, emotions and behaviour. However, the framework inadequately explains the impact of the social context of the customer experience.

Klaus and Maklan (2011) established five dimensions of customer experience in context the of sports tourism industry. These include hedonic enjoyment which is felt, perceived and experienced by participants at any time in the camp; social interaction refers to a sense of belonging to a community where members share common ground; efficiency enables the core experience for the customers, organizations and service providers; surreal feelings are described as recollection of a dream-like state, an almost altered stage of reality based on a certain activity and lastly, personal progression refers to making "progress" improving their free ride skills and techniques like goal, challenge, coaching, peer influence and self motivation. They remarked that improving customer experience is a growing priority for market research because experience is replacing quality as the competitive battleground for marketing (p. 2). Customer experience quality's core tenants as given by Klaus and Maklan (2011) include

1. It is assessed as an overall perception by customers and not as a gap to expectations.
2. Customers' assessment is based on overall value in use and not just a summation of performance during individual service episodes.
3. The measure of experience has a broader scope than that proposed by SERVQUAL. It includes emotions and peer influences.
4. Experience begins before service encounters and continues after the encounters.
5. Experience is assessed against service encounters across all channels.
6. An ideal measure should link more directly to customer behaviour and business performance than do either SERVQUAL or customer satisfaction.

Research with practitioners indicates that most firms use customer satisfaction, or its derivative Net Promoter Score, to assess their customers' experiences. Klaus and Maklan (2013) questioned this practice and proposed the principles of a new measure appropriate for the modern conceptualization of customer experience: the customer experience quality (EXQ) scale. In this chapter, we extend that work to financial service contexts to support a claim of generalizability better and compare its predictive power with that of customer satisfaction. We propose that EXQ better explains behavioural intention and recommendation than customer satisfaction.

Klaus and Maklan (2012, 2013) identified product experience, outcome focus, moment of truth and peace of mind as significant components of customer experience quality.

Product experience refers to the importance of customers' perception of having choices and the ability to compare offerings. (1) Outcome focus. It is associated with reducing customers' transaction cost, such as seeking out and qualifying new providers. (2) Moments of truth. It emphasizes the importance of service recovery and flexibility, dealing with customers once complications arise in the process of acquiring a service.

(3) Peace of mind. This dimension describes the customers assessment of all the interactions with the service providers before, during and after securing a

mortagage. This dimension includes statements strongly associated with the emotional aspects of service (Liljander and Strandvik 1997; Edvardsson 2005) and takes many items from the qualitatively generated dimension of provider experience. The dimension is reflecting the emotional benefits customers experience based on the perceieved expertise of the service provider and guidance throughout the process, which appeared to the customers not only as easy (Dabholkar et al. 1996), but also seemed to be "putting them at ease and subsequently, "increasing their confidence in the provider". Customers react to the peace of mind often with a notion of looking a building "a relationship" with a service provider rather than looking at the mortgage in a "purely transactional way" (Geyskens et al. 1996). Thus, it is important for companies to optimize the experience to ensure the continuity of customer relationship. However, to create an ultimate experience for the consumer, a company needs to understand factors influencing his or her decision-making and especially the motivation for initiating an act.

Although all customer experience conceptualizations are relevant, the present study is primarily conducted to validate the customer experience quality scale. Based on Klaus and Maklan (2013) we hypothesized that all four dimensions—peace of mind, moment of truth, outcome focus and moment of truth—contribute significantly to customer experience quality.

Hypothesis 1: Peace of mind, moment of truth, outcome focus and moment of truth contribute significantly to customer experience quality.

7.2.2 Customer Experience Outcomes

The service marketing literature on performance outcome measures such as consumer satisfaction, loyalty and brand equity is well established. Although extant literature indicates a strong connection between customer experience and customer satisfaction, empirical linkages between experience and performance measures are yet to be established. Kumar et al. (2007) discussed customer experience and remarked that customer expectations along with customer experience can create high satisfaction scores, whereas negative expectation and a positive experience towards the service create highest satisfaction scores as well. Kim et al. (2008) and Shankar et al. (2003) illustrated that overall, the more favourable the prior experience is, the higher is the satisfaction. To add, the relationship between service satisfaction and loyalty is nonlinear; that is, when satisfaction increases above a certain level, customer loyalty will increase rapidly (Oliva et al. 1992). Arnold et al. (2005) have also discussed outcomes of terrible experience. They remarked that customers who endured terrible experiences would either tell others about their bad experience, i.e. negative word-of-mouth, or voiced their complaints as a result of the terrible occurrences in the form of sending emails, writing letters, etc. or would stop visiting the provider again, i.e. discontinue patronage and switch to other.

So, customer experience not only drives customer satisfaction and loyalty intentions but also word-of-mouth (Keiningham et al. 2007). Subsequently, the relationship between customer experience and word-of-mouth is also being validated. Furthermore, the study proposed the stronger relationship between customer experience and service value which needs to be tested in the literature.

Klaus and Maklan (2013) remarked that to ensure satisfaction, positive word-of-mouth and loyalty, the whole experience should be well designed from the very first touch point that prospective customer may have with the company. The authors have established nomological validity of customer experience quality (CEX) with respect to customer satisfaction, word-of-mouth and loyalty intentions. Customer experience quality has significant impact on customer satisfaction, loyalty intentions, word-of-mouth and service value. Hence, we also hypothesized the following:

Hypothesis 2: Customer experience has a significant positive impact on customer satisfaction.
Hypothesis 3: Customer experience has a significant positive impact on loyalty intentions.
Hypothesis 4: Customer experience has a significant positive impact on word-of-mouth behaviour.
Hypothesis 5: Customer experience has a significant positive impact on service value.
Hypothesis 5: Customer experience has a stronger positive impact on loyalty intentions than customer satisfaction.
Hypothesis 6: Customer experience has a stronger positive impact on word-of-mouth behaviour than customer satisfaction.
Hypothesis 7: Customer experience has a stronger positive impact on service value than customer satisfaction.

7.3 Research Methodology

The Klaus and Maklan (2013) EXQ items are as such used to examine the reliability and validity of the EXQ scale in financial sector of emerging economies—India.

The respondents comprised of customers of Jammu City, India, who have experienced one of the three services, that is *Banking or Insurance or Investment* from the public and private sectors. The area sampling is used for the selection of customers from selected household. The Jammu City at the outset was divided into four geographically divided regions, that is north, south, east and west. One ward from each region was selected in the next stage, and later one locality from each ward from four regions (Gandhi Nagar, Trikuta Nagar, Krishna Nagar and Shastri Nagar) was selected. One customer availing services from last at least three years from each household for one of the three services was selected for the collection of data. The items generated for customer experience instrument are based on seven-point scale (1 = strongly disagree, 7 = strongly agree) or as Do not know/ Not applicable. The finalized customer experience scale comprised of 7 brand experience items change after, please, 11 of service experience, 05 belongs to post-purchase experience, 05 of customer satisfaction, 05 belongs to behavioural loyalty intentions, 07 of word-of-mouth and 07 of service value.

7.3.1   Descriptive Statistics

Before initiating data analysis, the descriptive analysis was undertaken to identify and delete non-normal items and outliers for proceeding for exploratory factor analysis. The items with skewness and kurtosis values greater than <0.5 were considered for deletion. A total of 21 items were deleted; that is, 03 items belonged to brand experience, 01 item was from service experience, 04 items belonged to post-purchase experience, 02 items were from behavioural loyalty intentions and customer satisfaction, 05 items belonged to word-of-mouth, and 04 from service value.

Constructs	Mean		Standard deviation		Skewness		Kurtosis	
	Min	Max	Min	Max	Min	Max	Min	Max
Peace of mind	3.07	3.18	1.31	1.38	0.04	0.07	0.8	1.2
Moment of truth	2.82	3.17	1.32	1.41	0.02	0.05	0.7	1.1
Outcome focus Experience	2.73	3.20	1.31	1.48	0.03	0.05	0.5	1.04
Product experience	*2.74*	*3.11*	*1.29*	*1.33*	*0.01*	*0.03*	*0.6*	*1.2*
Customer satisfaction	2.60	2.78	1.36	1.54	0.03	0.06	0.03	0.05
Behavioural loyalty Intentions	2.28	2.87	1.36	1.50	0.03	0.06	0.04	0.07
Word-of-mouth	2.58	2.84	1.11	1.30	0.01	0.03	0.03	0.08
Service value	2.58	2.72	1.09	1.39	0.01	0.03	0.04	0.07

7.3.2   Scale Development

The study validates the customer experience quality scale through confirmatory factor analysis (CFA).

7.3.3   Confirmatory Factor Analysis

The customer experience scale is tested individually using CFA initially, and its nomological validity is tested using SEM using marketing outcomes, that is customer satisfaction, behavioural loyalty intentions, word-of-mouth and service value. The results are discussed as under:

7.3.3.1 Customer Experience Quality

The validity of the customer experience quality (EXQ) scale as second-order scale is examined using SFA. The EXQ as a four-dimensional scale is confirmed; however, the model is found to be marginally fit. The chi-square, CMIN/df, CFI, RMSEA and NFI came to be 518.514, 3.457, 0.777, 0.081 and 0.719. All the four factors are found to be significant predictors of EXQ. All the items under these factors significantly contribute to their respective factors. The CR values were above the threshold criteria (greater than 1.96) and the SRW ranged between 0.55 and 0.98 (Table 7.1).

Table 7.1 Confirmatory factor analysis

Factor	Items	CR	SRW
Peace of mind	Confidence in the service provider expertise.	6.316	0.641
	Feeling with the service provider is easy	5.647	0.545
	Looking after the customer for a long time.	6.332	0.644
	The service provider gives independent advice on which product or service will best suit my needs	8.016	0.986
	I have built a personal relationship with the people at the service provider	–	0.553
	The service provider advises me throughout the service delivery process	6.567	0.681
Moment of truth	The service provider demonstrate flexibility in dealing with me	9.134	0.743
	The service provider will keep me up to date	9.852	0.797
	The service provider has a good reputation	9.588	0.777
	Service provider has good people skills.	9.011	0.734
	The service provider deal(t) well with me when things go wrong	–	0.742
Outcome focus	The service provider delivers a good customer service	6.672	0.662
	The service provider knows exactly what I want	8.064	0.886
	The service providers' facilities are better design to fulfil my needs than their competitors	7.320-	0.755
	The service providers' facilities are designed to be as efficient as possible for me		0.601
Product experience	The service provider offerings are superior to their competitors	4.828	0.629
	The service provider offerings have the best quality	2.919	0.624
	I do not choose the service provider by the price of his offerings alone	8.050	0.824
	At the service provider, I always deal with same forms and same people		0.558
Model fit indices		chi-square = 518.514 CMIN/df = 3.457 NFI = 0.719 CFI = 0.777 RMSEA = 0.081	

7.3.3.2 Customer Experience with Marketing Outcomes

To validate the results of EXQ, SEM is applied between EXQ and the four marketing outcomes—customer satisfaction, behavioural loyalty intentions, word-of-mouth and service value.

Customer Experience—Customer Satisfaction (Table 7.2)

The chi-square, CMIN/df, CFI, RMSEA and NFI values for the customer experience—customer satisfaction model came to be 728.199, 2.924, 0.812, 0.082 and 0.742 showing marginal model fitness. The SRW of the items lie between 0.55 and 0.91 with all critical ratio values above 1.96, indicating that all the items have significant impact.

Table 7.2 Confirmatory factor analysis—customer satisfaction

Factor	Items	CR	SRW
Peace of mind	I am confident in the service provider expertise	6.302	0.642
	Feeling with the service provider is easy	5.365	0.546
	The service provider will look after me for a long time	6.315	0.644
	The service provider gives independent advice on which product or service will best suit my needs	7.992	0.986
	I have built a personal relationship with the people at the service provider		0.552
	The service provider advises me throughout the service delivery process	6.548	0.681
	The service provider demonstrates flexibility in dealing with me	9.080	0.742
Moment of truth	The service provider will keep me up to date	9.805	0.796
	The service provider has a good reputation	9.541	0.776
	The people I am dealing with have good people skills	9.041	0.739
	The service provider deal(t) well with me when things go wrong		0.741
	The service provider delivers a good customer service	6.652	0.658
Outcome focus	The service provider knows exactly what I want	8.095	0.888
	The service providers' facilities are better design to fulfil my needs than their competitors	7.341	0.756
	The service providers' facilities are designed to be as efficient as possible for me		0.603
	The service provider offerings are superior to their competitors		0.637

(continued)

Table 7.2 (continued)

Factor	Items	CR	SRW
Product experience	The service provider offerings have the best quality	3.960	0.622
	I do not choose the service provider by the price of his offerings alone	3.900	0.818
	At the service provider, I always deal with same forms and same people	8.161	0.560
Customer satisfaction	My feelings towards the service provider are very positive		0.902
	I feel good about coming to the service provider for the offerings I am looking for	16.683	0.887
	Overall I am satisfied with the service provider and the service they provide	15.622	0.862
	I feel satisfied that the service provider produces the best results that can be achieved for me	18.140	0.917
	The extent to which the service provider has produced the best possible outcome for me is satisfying	16.331	0.879
Model fit indices		chi-square = 728.1999 CMIN/df = 2.924 NFI = 0.742 CFI = 0.812 RMSEA = 0.082	

Customer Experience—Behavioural Loyalty Intentions (Fig. 7.1)

The customer experience—behavioural loyalty intentions model indicate that the items were significantly and positively contributing and the SRW ranged between 0.50 and 0.98. All the critical ratio values showed significant results (Table 7.3).

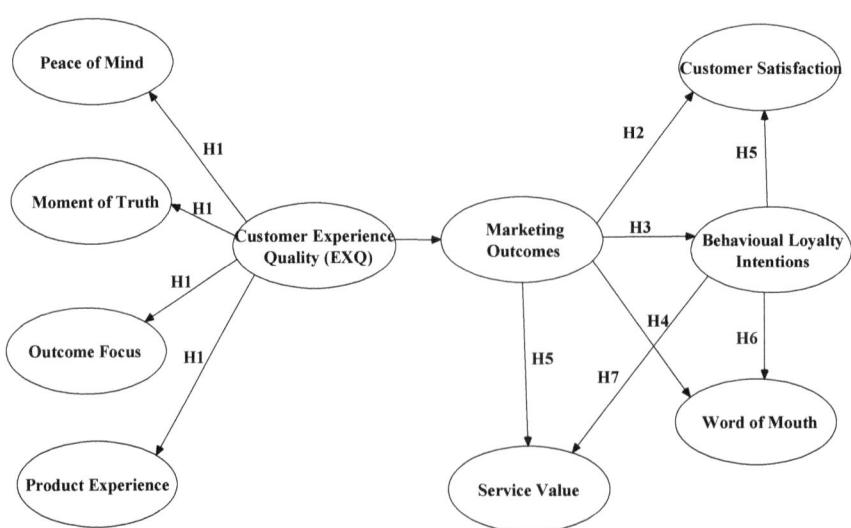

Fig. 7.1 Conceptual model of customer experience quality (EXQ)

Table 7.3 Confirmatory factor analysis—behavioural loyalty intentions

Factor	Items	CR	SRW
Peace of mind	I am confident in the service provider expertise	6.345	0.641
	Feeling with the service provider is easy	5.637	0.545
	The service provider will look after me for a long time	6.319	0.649
	The service provider gives independent advice on which product or service will best suit my needs	7.998	0.553
	I have built a personal relationship with the people at the service provider		0.681
	The service provider advises me throughout the service delivery process	6.553	0.742
	The service provider demonstrates flexibility in dealing with me	9.102	0.742
Moment of truth	The service provider will keep me up to date	9.835	0.797
	The service provider has a good reputation	9.575	0.777
	The people I am dealing with have good people skills	9.010	0.735
	The service provider deal(t) well with me when things go wrong		0.742
	The service provider delivers a good customer service	6.655	0.662
Outcome focus	The service provider knows exactly what I want	8.050	0.887
	The service providers' facilities are better design to fulfil my needs than their competitors	7.301	0.755
	The service providers' facilities are designed to be as efficient as possible for me		0.600
	The service provider offerings are superior to their competitors	3.462	0.632
Product experience	The service provider offerings have the best quality	2.914	0.642
	I do not choose the service provider by the price of his offerings alone	8.068	0.822
	At the service provider, I always deal with same forms and same people		0.558
Customer satisfaction	My feelings towards the service provider are very positive		0.500
	I feel good about coming to the service provider for the offerings I am looking for	4.856	0.522
	Overall I am satisfied with the service provider and the service they provide	6.273	0.863
	I feel satisfied that the service provider produces the best results that can be achieved for me	6.179	0.826
	The extent to which the service provider has produced the best possible outcome for me is satisfying	6.237	0.848
Model fit indices		chi-square = 799.984 CMIN/df = 3.213 NFI = 0.774 CFI = 0.747 RMSEA = 0.072	

Customer Experience—Word-of-Mouth

The model showed chi-square, CMIN/df, CFI, RMSEA and NFI values as 810.941, 2.981, 0.799, 0.775 and 0.075, respectively. All critical ratio values were greater than 1.96, and SRW was between 0.54 and 0.98 (Table 7.4).

Table 7.4 Confirmatory factor analysis— word-of-mouth behaviour

Factor	Items	CR	SRW
Peace of mind	I am confident in the service provider expertise	6.305	0.641
	Feeling with the service provider is easy	5.638	0.545
	The service provider will look after me for a long time	6.321	0.644
	The service provider gives independent advice on which product or service will best suit my needs	8.002	0.986
	I have built a personal relationship with the people at the service provider		0.553
	The service provider advises me throughout the service delivery process	6.556	0.681
	The service provider demonstrates flexibility in dealing with me	9.125	0.742
Moment of truth	The service provider will keep me up to date	9.850	0.797
	The service provider has a good reputation	9.591	0.777
	The people I am dealing with have good people skills	9.014	0.734
	The service provider deal(t) well with me when things go wrong		0.743
	The service provider delivers a good customer service	6.657	0.662
Outcome focus	The service provider knows exactly what I want	8.042	0.887
	The service providers' facilities are better design to fulfil my needs than their competitors	7.300	0.755
	The service providers' facilities are designed to be as efficient as possible for me		0.600
	The service provider offerings are superior to their competitors	6.294	0.630
Product experience	The service provider offerings have the best quality	6.912	0.642
	I do not choose the service provider by the price of his offerings alone	8.054	0.824
	At the service provider, I always deal with same forms and same people		0.558

(continued)

Table 7.4 (continued)

Factor	Items	CR	SRW
Customer satisfaction	My feelings towards the service provider are very positive		0.835
	I feel good about coming to the service provider for the offerings I am looking for	9.468	0.693
	Overall I am satisfied with the service provider and the service they provide	11.503	0.796
	I feel satisfied that the service provider produces the best results that can be achieved for me	13.637	0.891
	The extent to which the service provider has produced the best possible outcome for me is satisfying	10.401	0.742
Model fit indices		chi-square = 810.941 CMIN/df = 2.981 NFI = 0.799 CFI = 0.775 RMSEA = 0.075	

Customer Experience—Service Value (Table 7.5)

The customer experience—service value model indicated that the items were positively contributing as the SRW and SMC were quite acceptable as per the threshold criteria. All the critical ratio values showed significant results.

Table 7.5 Confirmatory factor analysis—service value

Factor	Items	CR	SRW
Peace of mind	I am confident in the service provider expertise	6.305	0.641
	Feeling with the service provider is easy	5.638	0.545
	The service provider will look after me for a long time	6.321	0.644
	The service provider gives independent advice on which product or service will best suit my needs	8.001	0.986
	I have built a personal relationship with the people at the service provider		0.553
	The service provider advises me throughout the service delivery process	6.556	0.681
	The service provider demonstrates flexibility in dealing with me	9.126	0.743

(continued)

Table 7.5 (continued)

Factor	Items	CR	SRW
Moment of truth	The service provider will keep me up to date	9.843	0.796
	The service provider has a good reputation	9.959	0.778
	The people I am dealing with have good people skills	9.014	0.734
	The service provider deal(t) well with me when things go wrong		0.743
	The service provider delivers a good customer service	6.660	0.633
Outcome focus	The service provider knows exactly what I want	8.043	0.886
	The service providers' facilities are better design to fulfil my needs than their competitors	7.302	0.755
	The service providers' facilities are designed to be as efficient as possible for me		0.600
	The service provider offerings are superior to their competitors	6.163	0.630
Product experience	The service provider offerings have the best quality	2.916	0.624
	I do not choose the service provider by the price of his offerings alone	8.059	0.823
	At the service provider, I always deal with same forms and same people		0.558
Customer satisfaction	My feelings towards the service provider are very positive		0.789
	I feel good about coming to the service provider for the offerings I am looking for	10.772	0.824
	Overall I am satisfied with the service provider and the service they provide	9.994	0.774
	I feel satisfied that the service provider produces the best results that can be achieved for me	8.022	0.642
	The extent to which the service provider has produced the best possible outcome for me is satisfying	9.568	0.746
Model fit indices		chi-square = 796.368 CMIN/df = 2.690 NFI = 0.793 CFI = 0.779 RMSEA = 0.081	

7.3.3.3   Overall Confirmatory Factor Analysis

The overall model that is customer experience with its four marketing outcomes namely customer satisfaction, behavioural loyalty intentions, word-of-mouth and service value showed chi-square (2190.927), CMIN/df (2.698), CFI (0.724), RMSEA (0.072) and NFI (0.789). All the customer experience values were significantly and positively contributing to the marketing outcomes (Table 7.6).

Table 7.6 Overall confirmatory factor analysis

Factor	Items	CR	SRW
Peace of mind	I am confident in the service provider expertise	6.304	0.641
	Feeling with the service provider is easy	5.637	0.545
	The service provider will look after me for a long time	6.320	0.644
	The service provider gives independent advice on which product or service will best suit my needs	8.000	0.986
	I have built a personal relationship with the people at the service provider		0.553
	The service provider advises me throughout the service delivery process	6.555	0.681
	The service provider demonstrates flexibility in dealing with me	9.126	0.742
Moment of truth	The service provider will keep me up to date	9.849	0.796
	The service provider has a good reputation	9.598	0.778
	The people I am dealing with have good people skills	9.020	0.734
	The service provider deal(t) well with me when things go wrong		0.743
	The service provider delivers a good customer service	6.657	0.633
Outcome focus	The service provider knows exactly what I want	8.039	0.886
	The service providers' facilities are better design to fulfil my needs than their competitors	7.298	0.755
	The service providers' facilities are designed to be as efficient as possible for me		0.600
	The service provider offerings are superior to their competitors	2.052	0.630

(continued)

Table 7.6 (continued)

Factor	Items	CR	SRW
Product experience	The service provider offerings have the best quality	2.916	0.624
	I do not choose the service provider by the price of his offerings alone	8.062	0.823
	At the service provider, I always deal with same forms and same people		0.558
Customer satisfaction	My feelings towards the service provider are very positive		0.569
	I feel good about coming to the service provider for the offerings I am looking for	4.481	0.590
	Overall I am satisfied with the service provider and the service they provide	5.879	0.878
	I feel satisfied that the service provider produces the best results that can be achieved for me	5.783	0.830
	The extent to which the service provider has produced the best possible outcome for me is satisfying	5.822	0.848
Model fit indices		chi-square = 2190.927 CMIN/df = 2.698 NFI = 0.789 CFI = 0.724 RMSEA = 0.072	

7.3.4 Reliability and Validity

The study assessed reliability and validity of the customer experience and its marketing outcomes by computing composite reliability and average variance extracted. The composite reliability of individual customer experience construct with the marketing outcomes came to be customer satisfaction (0.97), loyalty intentions (0.97), word-of-mouth (0.97) and service value (0.97). Further, convergent validity assumes that measures of construct should be theoretically related to each other and in practice as well. As the dimensions of customer experience have shown critical ratio values (more than 1.96) and standardized regression weight (more than 0.50) in CFA which established the convergent validity of the scale. In addition, convergent validity was also established by examining the average variance extracted (AVE) of each customer experience construct with the marketing outcomes. The AVE of the following constructs namely customer namely customer experience with customer satisfaction (0.60), loyalty intentions (0.60), word-of-mouth (0.62) and service value (0.61) are above the threshold criteria of 0.50 hence indicating convergent validity.

Dimensions	CR	AVE	α Value
Peace of mind (POM)	0.91	0.66	0.88
Moment of truth (MOT)	0.90	0.78	0.87
Outcome focus (OF)	0.92	0.71	0.81
Product experience (PE)	0.91	0.62	0.80
Customer satisfaction (CS)	0.97	0.60	0.95
Word of mouth (WM)	0.97	0.62	0.91
Behavioural loyalty intentions (BL)	0.97	0.60	0.84
Service value (SV)	0.97	0.61	0.87
Goodness of fit indices			
CMIN	df	CFI	IFI	RMSEA
2196	812	0.721	0.726	0.08

7.3.5   Results of Hypothesis Testing

The results for the customer experience quality, customer satisfaction, behavioural loyalty intentions, word-of-mouth and service value are illustrated in Table 7.3. The study shows that customer experience has a significant and positive impact on the customer satisfaction (0.52), behavioural loyalty intentions (0.79), word-of-mouth (0.57) and service value (0.50). Hence, in both the settings (UK and India), four hypotheses which relate to CE–CS (H1), CE–LI (H2), CE–WOM (H3) and CE–SV (H4) are accepted. Further, hypothesis (H5) that is related a stronger positive impact on loyalty intentions (0.765) than customer satisfaction (0.462) is accepted. While (H6 and H7) stronger impact on customer satisfaction in regard to word-of-mouth behaviour (0.449) and service value (0.699) are respectively, rejected.

Hypothesis 1	EXQ—Customer satisfaction	0.524
	PEA–EXQ	0.315
	MOM–EXQ	0.448
	OUT–EXQ	0.784
	PRO–EXQ	0.286
Hypothesis 2	EXQ—Behavioural loyalty intentions	0.788
	PEA–EXQ	0.297
	MOM–EXQ	0.597
	OUT–EXQ	0.558
	PRO–EXQ	0.267
Hypothesis 3	EXQ—Word-of-mouth	0.575
	PEA–EXQ	0.359
	MOM–EXQ	0.476
	OUT–EXQ	0.703
	PRO–EXQ	0.265
Hypothesis 4	EXQ—Service value	0.50
	PEA–EXQ	0.268
	MOM–EXQ	0.582
	OUT–EXQ	0.574
	PRO–EXQ	0.240
Hypothesis 5	EXQ—Customer satisfaction—Loyalty intentions	0.462–0.765
Hypothesis 6	EXQ—Customer satisfaction—Word-of-mouth	0.535–0.449
Hypothesis 7	EXQ—Customer satisfaction—Service value	0.769–0.699

Therefore, all results of the hypotheses are shown in Table 7.7.

Table 7.7 Hypotheses confirmation

Hypothesis	Customer experience has a significant positive impact upon…	Confirmation	Evidence (Path estimate scores)
1	Customer satisfaction	Yes	Path estimate of 0.524
2	Behavioural loyalty intentions	Yes	Path estimate of 0.788
3	Word-of-mouth	Yes	Path estimate of 0.575
4	Service value	Yes	Path estimate of 0.50
Customer experience has a higher positive impact than customer satisfaction upon …			
5	Behavioural loyalty intentions	Yes	0.462–0.765
6	Word-of-mouth	No	0.535–0.449
7	Service value	No	0.769–0.699

7.3.6 Discussion

The present study is an extended work on customer experience quality (EXQ) undertaken by Maklan and Klaus (Forthcoming) in service sector of UK setting. This study contributes to the extant marketing literature by validating the domain of consumer experience quality in financial service sector (banking, insurance and investment) operating in emerging economies, that is India, and its impact on marketing outcomes (customer satisfaction, behavioural loyalty intentions, word-of-mouth and service value).

The customer experience quality scale is assessed through validity and reliability analysis assuring the validation of EXQ scale in Indian settings. It is validated as four-dimensional scale, that is peace of mind, moment of truth, outcome focus and product experience. Also, the findings suggest that all the four individual dimensions (peace of mind, moment of truth, outcome focus and product experience) have positive and significant impact on the marketing outcomes, that is customer satisfaction, behavioural loyalty intentions, word-of-mouth and service value. Further, there is significant influence of each individual factor of customer experience quality on the marketing outcomes (Table 7.8). Moreover, there exist stronger and positive relationship between customer experience and behavioural loyalty intentions than customer satisfaction. The results put forth that a loyal customer is more likely to find the service encounter and the overall experience (with the service provider) more satisfying in comparison with non-loyal customer. Yi and Gong (2009) put forth that to improve customer satisfaction and loyalty, it is important to pay attention towards customer experience. In the context of UK services, the customer experience quality scale works well. Perhaps, the opinions of Indian customers in regard to EXQ were somewhat different. According to the Indian context, moment of truth (0.97) seems to be the most important factor that has to be considered by financial services for creating favourable customer experience quality followed by outcome focus (0.94), peace of mind (0.82) and product experience (0.74). The dimension product experience displays weakest association among all dimensions (outcome focus, peace of mind and moment of truth) with customer satisfaction (0.28), behavioural loyalty intentions (0.26), word-of-mouth (0.26) and service value (0.24). The relatively weak association with all marketing outcomes suggests that customer awareness about competitive services has increased and they no more accept every type of services from the same service provider because of varied customers' choices and their ability to compare offerings with different service providers. This gives customers the feeling of having a choice, and without a choice, they will unlikely to accept the offer no matter how good the offer was. So, it is important that all service providers should adopt appropriate strategy that can improve their services and increase customers' optimal experience which will subsequently enhance customer satisfaction, word-of-mouth, behavioural loyalty intentions and service value.

Moreover, the dimension *moments of truth* has strongest association with behavioural loyalty intentions and service value as it emphasizes on the importance

Table 7.8 Confirmatory factor analysis with individual marketing outcomes

Factor 1 = Customized services	CE–BL	0.674
	CE–CS	0.643
	CE–WM	0.604
	CE–SV	0.781
Factor 2 = Informative services	CE–BL	0.742
	CE–CS	0.647
	CE–WM	0.576
	CE–SV	0.716
Factor 3 = Reputation	CE–BL	0.678
	CE–CS	0.657
	CE–WM	0.596
	CE–SV	0.773
Factor 4 = Relationship commitment	CE–BL	0.641
	CE–CS	0.672
	CE–WM	0.574
	CE–SV	0.815
Factor 5 = Responsiveness	CE–BL	0.674
	CE–CS	0.677
	CE–WM	0.558
	CE–SV	0.788

Note CE customer experience, *BL* behavioural loyalty intentions, *CS* customer satisfaction, *WM* word-of-mouth and *SV* service value

of service recovery and flexibility in dealing with the customers wherein complications arise in the process of acquiring the services. Further, it is assumed that there is an importance of experiences related to the service company ("*service provider deal well with me when things go wrong*" *and* "*service provider has good people skills*") to form the positive behavioural intentions and influence the loyalty of the customers (Buttle and Burton 2002).

Lastly, the findings demonstrate greater influence of outcome focus on customer satisfaction and word-of-mouth. This strong association illustrates that in the Indian context, the customers are more satisfied with the service provider that gives accurate information of the services to the customers and also delivers good customer service. For instance, it reflects the importance of goal-oriented experiences such as providing a good match of service products to the customers. These goal-oriented experiences would endure customers to tell others about their experiences or may recommend the service provider to friends and family or would visit the service provider again.

From overall perspective, there exists stronger relationship between customer satisfaction (CE–CS) and customer experience quality than word-of-mouth (CE–WM) and service value (CE–SV). This indicates that even though customers are satisfied with the services but they do not prefer to tell to others about the services offered by the service providers to maintain their and the service providers confidentiality. In this context, Shankar et al. (2003) opined that overall the more favourable the experience is, the higher is the satisfaction. Similarly, Chen et al. (2008) considered actual experience and expectation as significant predictors of satisfaction. They remarked that the gap between expectation and actual experience leads to satisfaction/dissatisfaction. Further, customer experience not only drives customer satisfaction and loyalty intentions but also word-of-mouth (Keiningham et al. 2007) which may be positive or negative. The positive word-of-mouth endorses positive experience, satisfaction and loyalty, while negative word-of-mouth enhances switching intentions.

7.3.7 Implications, Limitations and Future Research

The present study of customer experience is validated by the findings which showed that there exists significant impact of customer experience on important marketing outcomes. The study illustrated a detailed structure whereby organizations can determine which dimensions of the customer experience are most strongly associated with the marketing outcomes. This allows the organizations to improve their customer experience management.

The research work to validate customer experience scale was conducted along the presence of certain unavoidable limitations. First, the study is based on three financial services such as banking, investment and insurance only, and for further research, it is suggested to adopt other services comprehensively to understand customer experience from their perspective. Secondly, the major limitation of the research is related to the presence of subjective responses of the customers with respect to customer experience constructs in the study. Nevertheless, appropriate efforts were taken to check the subjectiveness of the responses using various validity and reliability methods. Lastly, the scale needs to be tested across various states to generalize the results.

References

Adhikari, A., & Bhattacharya, S. (2015). Appraisal of literature on customer experience in tourism sector: Review and framework. *Current Issues in Tourism, 19*(4), 1–26.

Arnold, M. J., Kristy, E. R., Nicole, P., & Jason, E. L. (2005). Customer delight in a retail context: investigating delightful and terrible shopping experiences. *Journal of Business Research, 58,* 1132–1145.

Berry, L. L., Eilen, A. W., & Lewis, P. C. (2006). Services clues and customer assessment of the service experience: Lessons from marketing. *Academy of Management Perspectives, 20*(3), 43–57.

Berry, L. L., & Seltman, K. D. (2007). Building a strong services brand: Lessons from Mayo clinic. *Business Horizons , 50,* 199–209.

Biedenbach, G., & Marell, A. (2009). The impact of customer experience on brand equity in a business-to-business service setting. *Journal of Brand Management, 17*(6), 446–458.

Buttle, F., & Burton, J. (2002). Does service failure influence customer loyalty? *Journal of Consumer Behavior, 1*(3), 217–227.

Carbone, L. P., & Haeckel, S. H. (1994). Engineering customer experience. *Marketing Management, 3*(3), 8–19.

Chen, A. C. H., Chang, Y. H., & Fan, F. C. (2008). An empirical investigation of the relationship among service quality expectations, actual experience and its gap toward satisfaction. *Northeast Decision Sciences Institute Proceedings-March*, 28–20.

Chung, N., & Kwon, S. J. (2009). The effects of customers' mobile experience and technical support on the intention to use mobile banking. *Cyberpsychology and Behaviour, 12*(5), 539–543.

Dabholkar, P. A., Thorpe, D. I., & Rentz, J.O. (1996). A measure of service quality for retail stories:Scale development and validation. *Journal of the Academy of Marketing Science, 24*(1), 3–16.

Ding, D. X., Yang, H., & Verma, R. (2011). Customer experience in online financial service—A study of behavioural intentions for techno ready market segments. *Journal of Service Management, 22*(3), 344–366.

Edvardsson, B. (2005). Service quality: Beyond cognitive assessment. *Managing service quality,15*(2), 127–31.

Frow, P., & Payne, S. A. (2007). Toward the perfect customer experience. *Journal of Brand Management, 15*(2), 89–101.

Geyskens, I., Steenkamp, J. -B. E. M., Scheer, L. K., & Kumar, N. (1996). The effects of trust and interdependence on relationship commitment: A transatlantic study. *International Journal of Research in Marketing, 13*(4), 303–317.

Gilmore, J. H., & Pine, B. J. (2002). Customer experience places: The new offering frontier. *Strategy and Leadership, 30*, 4–11.

Grace, D., & O"Cass, A. (2004). Examining service experiences and post-consumption evaluations. *Journal of Services Marketing, 18*(6), 450–461.

Haeckel, S. H., Lewis, P. C., & Berry, L. L. (2003). How to lead the customer experience. *12*, 18–23.

Helkkula, A. (2011). Characterizing the concept of service experience. *Journal of Service Management, 22*(3), 367–389.

Keiningham, T., Cooli, B., Andreasseen, T., & Aksoy, L. (2007). A longitudinal examination of Net Promoter and firm revenue growth. *Journal of Marketing, 7*, 139–151.

Kim, S. H., Cha, J., Knutson, B. J., & Beck, J. A. (2011). Development and testing of the consumer experience index (CEI). *Managing Service Quality, 21*(2), 112–132.

Kim, K. H., Kim, S. K., Kim, D. Y., Kim, J. H., & Kang, S. H. (2008). Brand equity in hospital marketing. *Journal of Business Research, 61*, 75–82.

Klaus, P., & Maklan, S. (2011). Bridging the gap for destination extremes sports: a model of sports tourism customer experience. *Journal of Marketing Management, 27*, 1–24.

Klaus, P., & Maklan, S. (2012). EXQ: A multiple-scale for assessing service experience. *Journal of Service Management, 23*(1), 5–33.

Klaus, P., & Maklan, S. (2013). Towards a better measure of customer experience. *International Journal of Market Research, 55*(2), 227–246.

Knutson, B., Beck, J., Kim, S., & Cha, J. (2006). Identifying the dimensions of the experience constructs. *Journal of Hospitality & Leisure Marketing, 15*(3), 31–47.

Kumar, R. N., Kirking, D. M., Hass, S. L., Vinokur, A. D., Taylor, S. D., Atkinson, M. J., & Mckercher, P. L. (2007). The association of consumer expectations, experiences and satisfaction with newly prescribed medications. *Quality Of Life Research, 16*(7), 1127–1136.

Liljander, V. & Strandvik, T. (1997). Emotions in service satisfaction. *International Journal of Service Industry Management, 8*(2),148–69.

Meyer, C., & Schwager, A. (2007). Understanding customer experience. *Harvard Business Review, 85*(2), 16–26.

Nagasawa, S. (2008). Customer experience management. *The TQM Journal, 20*(4), 312–323.

Oliva, A. T., Oliver, L. R., & Mac Millan, I. C. (1992). A catastrophe model for developing service satisfaction strategies. *The Journal of Marketing, 56*(3), 83–95.

Pareigis, J., Edvardsson, B., & Enquist, B. (2011). Exploring the role of the service environment in forming customer's service experience. *International Journal of Quality and Service Sciences, 3*(1), 110–124.

Payne, A. F., Storbacka, K., & Frow, P. (2008). Managing the co-creation of value. *Academy of Marketing Science, 36,* 83–96.

Pine II, B. J., & Gilmore, J. H. (1999). The Experience Economy. *Harvard Business Review,* July-August, Boston, M.A, 18–23.

Rias, N. M., Musab, R., & Muda, M. (2016). Reconceptualisation of customer experience quality (EXQ) measurement scale. *Procedia Economics and Finance, 37,* 299–303.

Schmitt, B. (1999). Experiential marketing: How to get customers to sense, feel, think, act and relate to your company and brands. *Journal of Marketing, 15*(1–3), 56–67.

Schmitt, B. (2000). Experiential marketing: How to get customers to sense, feel, think, act and relate to your company and brands. *Journal of Marketing Research, 30,* 7–27.

Shankar, V., Smith, A. K., & Rangaswamy, A. (2003). Customer Satisfaction and loyalty in online and offline environments. *International Journal of Research in Marketing,* 153–175.

Terblanche, N. S. & Boshoff, C. (2004). The in-store shopping experience: A comparative study of supermarket and clothing store customers. *South African Journal of Business Management, 35*(4), 1–10.

Verhoef, P. C., Lemon, K. N., Parasuraman, A., Roggeveen, A., Tsiros, M., & Schlesinger, L. A. (2009). Customer experience creation: Determinants, dynamics, and management strategies. *Journal of Retailing, 85*(1), 31–41.

Yi, Y., & Gong, T. (2009). An integrated model of customer social exchange relationship: The moderating role of customer experience. *The Service Industries Journal, 29*(11), 1513–1528.

Chapter 8
Re-investigating Market Orientation and Environmental Turbulence in Marketing Capability and Business Performance Linkage: A Structural Approach

Jagmeet Kaur, Hardeep Chahal and Mahesh Gupta

Abstract This study aims to re-investigate the role of market orientation as an antecedent to marketing capability and effect of environmental turbulence as moderating variable in marketing capability—competitive advantage—business performance relationship. Data are collected from multiple respondents, that is, a branch manager and three senior managers of 144 branches of public and private banks operating in Jammu city, North India. The study establishes marketing capability as a three-dimensional construct, comprising outside-in, inside-out and spanning, unlike majority of previous studies on marketing capability. The study also supports the school of thought which believes that market orientation acts as an antecedent to marketing capability rather than its dimension. Further, the findings reveal partial mediating role of marketing capability on market orientation and competitive advantage linkage. However, environmental turbulence does not moderate in marketing capability-competitive advantage relationship in the tech-savvy banking sector.

Keywords Business performance · Competitive advantage · Environmental turbulence · Market orientation and marketing capability

J. Kaur
Government General Zorawar Singh Memorial Degree College, Reasi, India
e-mail: kaurjagmeet9@gmail.com

H. Chahal (✉)
Department of Commerce, University of Jammu, Jammu, India
e-mail: chahalhardeep@rediffmail.com

M. Gupta
University of Louisville, Louisville, USA
e-mail: Mahesh.gupta@louisville.edu

8.1 Introduction

Marketing scholars such as Day (1994), Vorhies and Morgan (2005) underscore that an organisation needs to build resources that are valuable, non-substitutable and inimitable to sustain competitive advantage (CA). They further remark that marketing capability (MC) is one of the market-based resources which facilitates the organisation to better understand market needs and develops long-term relationship with different stakeholders. Day (1994) also puts forth that an organisation with marketing capability is able to perform varied marketing activities such as market information generation market positioning, segmentation, market planning activities, product development activities and promotion efficiently. In 2011, Day explicates that marketing capability, by minimising the gap between market complexities and organisational ability, enables a firm to cope up with market changes.

Evidence within marketing literature suggests that marketing capability leads to achieve superior business performance, and hence, the concept requires more extensive attention (Day 1994, 2011; Hooley et al. 1999; Vorhies and Morgan 2005; Pham et al. 2017). Further, the researchers recommended that there is a need to explore the impact of antecedents, such as market orientation, differentiation strategies, organisational culture, niche versus broad market selection, industry context, learning, on marketing capability (Vorhies 1998; Vorhies and Yarbrough 1998). Vorhies (1998), Vorhies and Yarbrough (1998) and Ros et al. (2010) also underline that the studies on marketing capability, competitive advantage and business performance relationship are scarce. For instance, Vorhies (1998) quotes that "research is needed that investigates how various marketing capability marketing capability contribute individually to organisational success" (p. 18). In line with this, Krasnikov and Jayachandran (2008) also pointed out that empirical research that examines the impact of marketing capability on business performance is scarce as compared to the impact of general capabilities such as R&D, operations in other functional areas. Recently, Pham et al. (2017) put forth that there is a need of examining and validating the role of marketing capability such as product development, pricing, marketing communication, distribution, in enhancing business performance in emerging economies. The study observed mixed relationships between marketing capability and business performance in Vietnam export sector. Marketing capability such as product development, pricing, marketing communication showed direct and positive relationship with business performance, whereas the relation of distribution and after-sale service was found to be insignificant. Lindblom et al. (2008) conducted study in retail sector and identified weak linkage of marketing capability and business performance. The scholars advocated that if variables such as competitive intensity, store location are considered in the marketing capability and business performance relationship, then such weak linkage can be controlled and strengthened as well. Hence, they suggested that the effects of the moderating factors on marketing capability–business performance relationship are required to be undertaken in the future. In this context, Najafi Tavani et al. (2016)

also highlighted the need of measuring role of environmental turbulence as a moderator in marketing capability–business performance relationship.

This study is conducted to identify the influence of market orientation (antecedent) on marketing capability and its subsequent impact on competitive advantage and business performance in service sector. Besides, the study also aims to re-investigate the role of market orientation as an antecedent to marketing capability, environment turbulence as a moderator in marketing capability, competitive advantage and business performance relationship. In addition, the role of competitive advantage as a mediator on marketing capability and business performance relationship would also be identified.

The structure of the paper is as follows: firstly, literature review and hypotheses development are presented. After that, discussion on hypotheses development and research methods used to test these hypotheses along with the results of hypotheses testing are addressed. The study ends with discussions, limitations and directions for future research.

8.2 Literature Review and Hypotheses Development

8.2.1 Marketing Capability

The marketing literature offers four different perspectives of marketing capability which include operational (Day 1994, 2011; Cadogan et al. 2002; Greenley et al. 2005), marketing mix (Vorhies and Yarbrough 1998; Vorhies and Harker 2000; Weerawardena 2003; Vorhies and Morgan 2005), intellectual capital (Moller and Antilla 1987) and competition (Fahy et al. 2000).

The operational perspective conceptualises marketing capability as the combination of outside-in, inside-out and spanning capability (Day 1994). This concept has been considered in the present paper. Researchers such as Vorhies (1998), Vorhies and Morgan (2005) focus on marketing mix elements such as 4Ps, marketing planning and implementation, market research to comprehend marketing capability. According to Moller and Antilla (1987), intellectual capital, which is the third perspective, considers marketing capability as the composition of human, market and organisational assets. Lastly, competition-based perspective of marketing capability puts emphasis on (Fahy et al. 2000) market orientation, time horizon of strategic decisions and positioning skills.

Regarding different conceptualisations of marketing capability, Guenzi and Troilo (2007) suggested that all these conceptualisations of marketing capability primarily focus on developing business capabilities to identify, anticipate and understand market requirements. While conceptualising marketing capability, Day (1994) remarks that though it is not possible to list all organisational capabilities but capabilities that correspond to core processes for creating business value can be identified.

This study focuses on operational perspective of marketing capability. From operational perspective, marketing capability can be conceptualised as an integration of firm's resources and employees' knowledge and skills to create superior customer value. Day (1994) and Tsai and Shih (2004) consider marketing capability as valuable, rare and inimitable which is difficult for the organisation to replace with any other resources. Day (1994) categorised marketing capability into outside-in, inside-out and spanning capabilities. Outside-in capability (such as market sensing, customer linking) helps an organisation in identifying and understanding the market needs (Day 1994; Greenley et al. 2005). Inside-out capability is internally focused marketing capability which is created to enhance firm's operational performance (Vijande et al. 2012). Lastly, spanning capability is the integration of outside-in and inside-out capabilities.

8.2.2 Market Orientation

Body of researchers such as Vorhies et al. (1999), Vorhies and Harker (2000), Hooley et al. (2005), Murray et al. (2011), Merrilees et al. (2011) and Ngo and O'Cass (2012) underlined market orientation as a significant antecedent of marketing capability. Morgan et al. (2009) put forth that market orientation, being market-based knowledge asset, can enhance marketing capability of an organisation by considering customers' needs, competitive strategy, channel requirement, etc., important for sustaining competitive advantage. However, Hooley et al. (1999) and Fahy et al. (2000) considered market orientation as one of the components of marketing capability. They remarked that market orientation as the dimension of marketing capability is the main strategy which facilitates an organisation in achieving superior performance. In this study, like other school of thought we argue that as an antecedent, market orientation provides a knowledge base that aid in the creation of valuable and inimitable marketing capability. Besides market orientation, there are other antecedents such as marketing knowledge management, market information processing capabilities, task routinisation, entrepreneurial intensity, management capabilities, organisational culture and structure considered in the literature that affects marketing capability. These antecedents, however, are associated either directly or indirectly with market orientation. For instance, marketing knowledge management (Tsai and Shih 2004) and market information processing (Vorhies 1998) are related with generation, dissemination and communication of information within and between departments and can be judged through market orientation. Further, entrepreneurial intensity (comprising innovative, proactive and risk-taking abilities of top management) and management capabilities (market support capabilities) are also indirectly linked with market orientation, since the availability of such skills in an organisation will enhance market orientation and its marketing capability.

However, Lafferty and Hult (2001) categorised the concept of market orientation into five perspectives. These perspectives include decision-making, strategic,

customer orientation, market intelligence and culturally based behavioural perspectives. The first perspective, that is, decision-making, revolves around the
commitment of top management in pursuing decision-making practices as well as
sharing of information among the employees. The strategic, second perspective, of
market orientation concentrates on strategy development and execution, while the
third perspective customer orientation focuses on developing a culture based on
customer orientation. The fourth perspective, market intelligence or behavioural
perspective, considers market orientation as the composite of market intelligence
generation, dissemination and responsiveness (Jaworski and Kohli 1993). Finally,
culturally based behavioural, the fifth perspective conceptualises market orientation
as an organisational culture which consists of three behaviour components which
include customer orientation, competitor orientation and inter-functional coordination (Narver and Slater 1990).

Lafferty and Hult further suggested although the characteristics of varied conceptualisations of market orientation available in the literature are different, all of
them agree that market orientation focuses on customers and gives importance to
sharing of information, inter-functional coordination of and responding to marketing activities by taking the appropriate action. Even though five different perspectives of market orientation exist in the literature, the study considers
behavioural perspective of market orientation (Fig. 8.1). Number of marketing
scholars such as Vorhies et al. (1999), Vorhies and Harker (2000), Morgan et al.
(2009), Murray et al. (2011) used behavioural perspective of market orientation to
determine its relationship with marketing capability as this perspective specifically
emphasises on organisation-wide usage of market information. Market-oriented
organisation constantly gathers information about current and future customers,

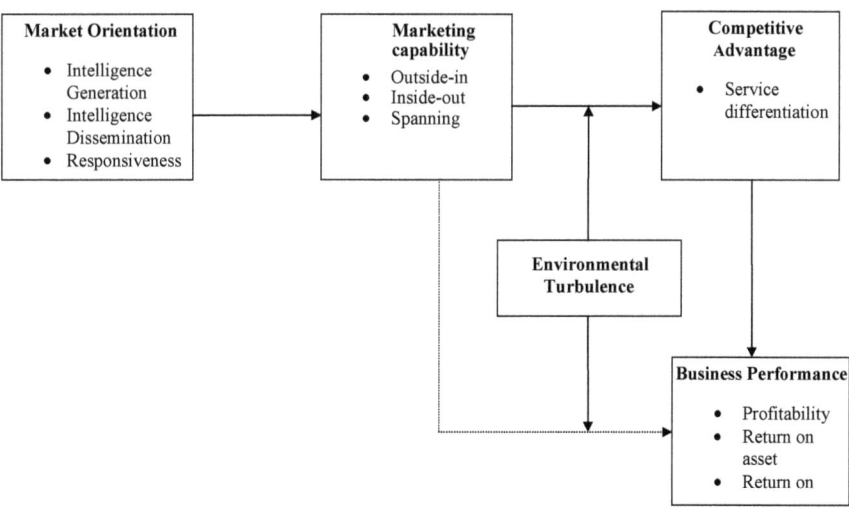

Fig. 8.1 Relationships of market orientation, marketing capability, competitive advantage,
environmental turbulence and business performance

competitor's capabilities and other market pressures such as technological changes, government policies (Murray et al. 2011). All these information are consequently shared among the employees of an organisation who quickly respond to these market changes. Thus, market orientation provides unique know-what knowledge base for the development of marketing capability which, in turn, helps the organisation in better serving target market (Morgan et al. 2009; Ngo and O'Cass 2012).

Market orientation refers to the organisation-wide generation of market intelligence about current and future customer needs, dissemination of its intelligence across departments and organisation-wide responsiveness to the disseminated intelligence (Jaworski and Kohli 1993). This perspective of market orientation is well accepted in the marketing capability literature, as it is considered as market-based knowledge asset which focuses on organisational and human activities pertaining to the creation, dissemination and reaction to market intelligence that may be difficult for managers to directly observe in their competitors.

Researchers observed that market orientation have positive and significant influence on marketing capability (Murray et al. 2011; Merrilees et al. 2011; Ngo and O'Cass 2012). They further remarked that since market-oriented organisation emphasises on meeting customers' needs as well as on building relationship with them, such organisation possesses better marketing capability in comparison to non-market-oriented organisation. Further, Qureshi and Mian (2010) contended that a market-oriented organisation is actively engaged in acquiring information to enhance its ability to predict prevailing market trends and also to unveil latent needs of the customers. Organisation which is first to uncover such latent needs is in a better position to develop the marketing capability. Hence, the study hypothesises that:

*H*1: Market orientation is a strong and positive predictor of marketing capability.

Morgan et al. (2009) express that market orientation-based marketing capability helps an organisation in resource adaptation better than the competitors. Murray et al. (2011) also observed that market orientation improves business performance through marketing capability. Therefore, while analysing market orientation–business performance linkage, the mediating impact of marketing capability is significant. Hence, we hypothesise that:

*H*2: Market orientation has significant and positive impact on (a) competitive advantage and (b) business performance through marketing capability.

The marketing literature confirms that marketing capability paves ways for achieving competitive advantage (Day 1994; Weerawardena 2003; Kaleka and Morgan 2017). Weerawardena (2003) puts forth that an organisation should keep on developing and nurturing marketing capability up to that point until they lead to achieve competitive advantage. Recently, Kaleka and Morgan (2017) observed that an organisation with marketing capability (i.e. informational, customer relationship and product development) is more competitive in attaining marketing differentiation. They also put forth that customer relationship capabilities create superior customer value which acts as a tool for gaining significant market information. Such information ultimately enhances the organisational ability to differentiate its product/service. Thus, it is hypothesised that:

*H*3: Marketing capability is positively and significantly related to competitive advantage.

Najafi Tavani et al. (2016) underlined that an organisation with marketing capability is able to build effective strategies and policies and hence enjoys superior business performance. Researchers such as Nath et al. (2010) and Krasniko demonstrated that organisation's financial performance is more significantly influenced by marketing capability than other functional capabilities such as operations capabilities, R&D. Krasniko and Jayachandran (2008) also remarked that marketing capability is in fact "success-producing capabilities" that keep an organisation ahead of its competitors, whereas R&D and operations capabilities can be seen as "failure prevention capabilities" which are essential for its survival. Azizi et al. (2009) also observed positive relationships of marketing capability with financial performances. They remark that marketing capability leads to develop effective marketing techniques in an organisation, and hence, this results in better performance. Thus, we hypothesise that:

*H*4: There is a strong relationship between marketing capability and business performance.

Vorhies and Yarbrough (1998), Murray et al. (2011) also observed indirect role of competitive advantage in marketing capability–business performance linkage. The study results of Vorhies and Yarbrough (1998) found that an organisation with marketing capability is comparatively better in attaining market growth, position and ROA. Further literature also discusses that strong relationship between marketing capability and business performance through competitive advantage. Murray et al. (2011) established the role of cost and differentiation strategies in improving marketing capability–business performance relationship. Specifically, the authors remarked that marketing capability such as pricing capability, marketing communication capability consequently creates differentiated image and leads to competitive advantage and business performance. Thus, it is hypothesised that:

*H*5: Marketing capability has positive and significant impact on business performance through competitive advantage.

The literature highlights that an organisation which develops marketing capability is able to respond to the environment that possesses competitive intensity, market and technological turbulences and hence, is effective in reducing uncertainty levels regarding environmental impacts better than the competitors (Vorhies and Yarbrough 1998). The researchers further assert that marketing capability facilitates an organisation to collect, process, distribute and react to environmental information better than the competitors which, in turn, paves way for achieving superior business performance. Hence, we hypothesise that:

*H*6: Environmental turbulence moderates the relationship between marketing capability and (a) competitive advantage and (b) business performance.

8.3   Research Methodology

8.3.1   Generation of Scale Items

Marketing capability is measured by using MARKCAPB scale developed by Chahal and Kaur (2014). Further, competitive advantage is assessed using self-developed items of service differentiation as the items for measuring the same in banking sector are not available in the literature. Jaworski and Kohli (1993) scale is used to measure market orientation and environmental turbulence. Lastly, business performance is assessed through financial indicators, namely market share growth, ROI, ROA and profitability (Vorhies and Harker 2000; Morgan et al. 2009).

8.3.2   Data Collection

The study is conducted in urban Indian banking sector operating in Jammu (i.e. 144 branches of twenty-one public and seven private banks). The data are collected from multiple respondents, that is, four (one branch manager and three senior managers/officers) from each bank's branch to minimise common method biasness. The response rate of the study was 52.60% which includes 303 fully filled questionnaires out of 576.

8.3.3   Exploratory Factor Analysis (EFA)

Before performing EFA, data were made normal by deleting sixteen outliers. The descriptive statistics of retained items is given in Table 8.1. Further, alpha values of the selected constructs were found to be above threshold value of 0.70.

EFA based on principal component analysis with a varimax rotation was conducted on 287 observations. All the items were examined for retention based on criteria such as eigenvalues (≥ 1) Kaiser–Meyer–Olkin (KMO) value (≥ 0.50), measure of sampling adequacy (MSA) (≥ 0.70), communality (≥ 0.50) and minimum factor loading (≥ 0.40) (Hair et al. 2009).

The study considers marketing capability as a three-dimensional construct. EFA was performed separately on all three dimensions of marketing capability, namely outside-in, inside-out and spanning. The EFA results are shown in Table 8.2.

Table 8.1 Descriptive statistics

Constructs	Mean		Standard deviation		Skewness		Kurtosis	
	Min	Max	Min	Max	Min	Max	Min	Max
MC	2.29	4.37	0.63	1.19	−0.02	−1.70	0.10	5.74
MO	3.26	4.28	0.60	1.14	−0.38	−1.44	−0.03	2.94
ET	3.62	4.15	0.55	0.88	−0.01	−1.11	0.18	0.83
CA	3.34	4.26	0.69	1.12	−0.42	−0.42	−0.02	2.47

Table 8.2 EFA results of marketing capability

	Dimensions	Factor loading	KMO	Variance (%)	Communality
	Outside-in		*0.76*	*67.72*	
F1: Relationship	Relationship with channel member	0.86		38.25	0.77
	Relationship with channel member w.r.t cash transfer	0.86			0.76
	Relationship with channel member w.r.t cheque processing	0.73			0.62
F2: Regularity	Regular customer contact	0.84		16.92	0.72
	Periodic market research	0.75			0.62
	Regular interdepartmental meetings	0.68			0.52
F3: Communication	Communicating service changes	0.84		12.55	0.74
	Effective solutions to customer problems	0.78			0.67
	Inside-out		*0.79*	*62.20*	
F1: Web technology	Product/service information on website	0.79		33.54	0.65
	Online relationship building	0.80			0.67
	User-friendly website	0.79			0.63
	Investment to upgrade technology	0.60			0.47
F2: Employee bonding	Encouraging employees	0.72		28.64	0.56
	Fair appraisal system	0.82			0.67
	Providing constant guidance to employees	0.81			0.71

(continued)

Table 8.2 (continued)

	Dimensions	Factor loading	KMO	Variance (%)	Communality
	Spanning		*0.71*	*74.67*	
F1: Effective brand and advertising	Developing and executing advertising programmes	0.87		26.52	0.77
	Managing advertising programmes	0.890			0.81
	Effective use of brand management	0.71			0.57
F2: Pricing skill	Using pricing skills	0.85		26.40	0.75
	Information about competitors pricing tactics	0.85			0.75
	Responding to competitors pricing tactics	0.73			0.66
F3: Product/ service skill	Developing new product/ service	0.89		21.75	0.84
	Launching new product/ service	0.86			0.84

The study identified three dimensions of outside-in (i.e. relationship, regularity and communication), two dimensions of inside-out (i.e. Web technology and employee bonding) and three of spanning, that is, effective brand and advertising; pricing and product/service skills were related to effective brand and advertising; pricing and product/service skills.

Similarly, EFA results identified four dimensions each of market orientation (intelligence generation-I, intelligence generation-II, intelligence dissemination and responsiveness) and competitive advantage (online bank services, cash/fund processing time, ATMs' service quality and draft/cheque processing) (Table 8.3). While five factors are revealed for environmental turbulence which include market turbulence-I (Marketing practices), market turbulence-II (Product/Service preference), competitive intensity, technological turbulence-I (Change in technology) and technological turbulence-II (Technological competitiveness) in the study.

8.3.4 Confirmatory Factor Analysis (CFA)

CFA was employed on selected study constructs—market orientation, marketing capability, environmental turbulence and competitive advantage—o assess and validate the measurement of factors included in the respective models. The CFA models were evaluated on the basis of fitness given in Table 8.4. Moreover, path value criteria, that is, critical ratio (CR) and standardised regression weight (SRW), are used to ensure that the observed variables load as predicted on the expected

Table 8.3 EFA results of MO, ET and CA

	Dimensions	Factor loading	KMO	Variance (%)	Communality
	Market orientation		0.82	64.08	
F1: Intelligence generation-I (Customer need-focused)	Customer's requirements are identified regularly	0.54		17.26	0.59
	Customer preferences change a lot over a short period of time	0.79			0.64
	Assessing impact of product prices on customers' expectation	0.70			0.64
	Assessing customer's opinion regarding product/services	0.53			0.51
F2: Intelligence generation-II (Customer satisfaction-focused)	Generating information about forces influencing customers' preferences	0.58		16.33	0.52
	Regular measures of improving customer service	0.87			0.78
	Measuring customer satisfaction on regular basis	0.83			0.77
F3: Intelligence dissemination	Interdepartmental meetings for discussing customer needs	0.66		16.02	0.53
	Periodic circulation of documents	0.78			0.67
	Frequent circulation of customer information	0.85			0.76
F4: Responsiveness	Business plans are driven more by market advances	0.56		14.46	0.49
	Taking corrective actions when customers are dissatisfied	0.750			0.73
	Modifying products/ services as per customer's desire	0.76			0.68

(continued)

Table 8.3 (continued)

	Dimensions	Factor loading	KMO	Variance (%)	Communality
	Environmental turbulence		*0.69*	*68.64*	
F1: Competitive intensity	Cutthroat competition	0.80		15.84	0.69
	Many promotion wars	0.79			0.69
	Anything that a competitor can offer other can match readily	0.67			0.56
F2: Market turbulence-I (Marketing practices)	Changes in marketing practices to keep up with competitors	0.88		14.09	0.85
	Changes in marketing practices to keep up with customers	0.87			0.81
F3: Technological turbulence-I (Change in technology)	Technological development is relatively minor	0.69		13.05	0.55
	Technology is not subject to very much change	0.74			0.58
	Difficult to forecast the technology	0.66			0.58
F4: Technological turbulence-II (Competitiveness)	New ideas are made through technological breakthrough	0.83		12.91	0.75
	Technological changes provide big opportunities	0.79			0.67
F5: Market turbulence-II (Product preference)	Frequent changes in customers product/ service preferences	0.82		12.75	0.73
	Customers tend to look new product all the time	0.83			0.77
	Competitive advantage		*0.82*	*75.42*	
F1: Online bank services	Instant access to account information on the Internet	0.79		20.89	0.72
	Instant access to loan statement on the Internet	0.85			0.76
	Quick downloading of pages from the Internet	0.82			0.77
	Secured online bill payment services	0.68			0.53

(continued)

Table 8.3   (continued)

	Dimensions	Factor loading	KMO	Variance (%)	Communality
F2: Cash/fund processing time	Processing time for depositing cash	0.89		20.22	0.87
	Processing time for cash withdrawal	0.90			0.88
	Quick fund transfer services	0.83			0.77
F3: ATMs service quality	ATMs are always operational	0.78		18.25	0.69
	Neat and clean surrounding around ATMs	0.88			0.85
	Secured ATMs services	0.87			0.83
F4: Draft/cheque processing	Draft delivery	0.87		16.07	0.81
	Clearance of cheque of bank's branch	0.85			0.81
	Clearance of cheque of other bank	0.59			0.52

number of factors. The critical ratio is the value of a test statistic which indicates a specified significance level. Values greater than 1.96 which denotes an estimate that is statistically significantly different from zero at the 0.05 level are used to retain the items. Similarly, standardised regression weight reflects the change in the dependent variable for each unit change in the independent variable (Hair et al. 2009). The SRW value less than 0.50 is considered to delete the items. However, utmost care is taken not to remove those items which are significant for the study.

Following EFA, CFA was run on identified factors of all the four constructs, namely marketing capabilities, market orientation, environmental turbulence and competitive advantage. The CFA results established marketing capability as multi-dimensional construct consisting of outside-in, inside-out and spanning capabilities. One item of spanning capabilities was deleted as the omission of the item resulted into better fit of the model.

Market orientation, environmental turbulence and competitive advantage were observed as multi-dimensional second-order constructs. In this stage, two items of market orientation and three of environmental turbulence are deleted due to low standardised regression weights. The standardised regression weights of all the items of the four constructs were greater than threshold criterion 0.50.

8.3.5   Common Method Bias

Common method variance is a form of systematic error variance which can cause observed correlations among variables to differ from their population values

Table 8.4 Measurement model

Constructs	χ^2/df	Normed fit index	Relative fit index	Incremental fit index	Tucker–Lewis index	Comparative fit index	Root mean square of error approximation
	Fit indices						
	<5	≥0.9	≥0.9	≥0.9	≥0.9	≥0.9	<0.08
Outside-in	2.10	0.94	0.89	0.97	0.95	0.97	0.06
Inside-out	2.13	0.95	0.93	0.98	0.96	0.98	0.06
Spanning	2.09	0.96	0.94	0.98	0.97	0.98	0.06
Marketing capability	2.61	0.79	0.76	0.86	0.84	0.86	0.07
Market orientation	2.71	0.88	0.84	0.92	0.88	0.92	0.07
Competitive advantage	2.51	0.93	0.91	0.96	0.94	0.96	0.07
Environmental turbulence	2.62	0.91	0.85	0.94	0.90	0.94	0.07

(Meade et al. 2007). It is a potential source of measurement error that can create a serious threat for the validity of conclusions about the associations among variables (Podsakoff et al. 2003). Hence, in order to avoid this problem, the study assesses the common method variance for all the constructs. First, we conducted Harman single factor test in which all variables of the constructs in the study are loaded into an exploratory factor analysis to determine the number of factors that are necessary to account for the variance. Podsakoff et al. (2003, p. 889) quote that "*the basic assumption of this technique is that if a substantial amount of common method variance is present, either (a) a single factor will emerge from the factor analysis or (b) one general factor will account for the majority of the covariance among the variables.*" Hence, to examine common method bias using this method, all variables of the marketing capability, market orientation, environmental turbulence and competitive advantage were entered into an exploratory factor analysis, using unrotated principal component factor analysis and principal component analysis with varimax rotation. The factor analysis revealed the presence of 16 distinct factors rather than a single factor. The 16 factors accounted for 69.61% of variance in which first largest factor accounted for 5.94% variance. The result indicates that no single factor emerged from this analysis which accounted for the majority of variance. Further, to cross-validate the results obtained from the above method, common latent factor was also performed. In this method, relationships between latent factor (created) and variables were developed in CFA to assess the variance explained by different relationships. The variance values obtained were less than 25% for marketing capability (10.89%), market orientation (6.25%), competitive advantage (20.25%) and environmental turbulence (6.25%). Thus, the results from both the methods denote that common method bias is not the subject of any concern in the study.

8.3.6 Reliability and Validity

The study performed psychometric analyses to provide evidence of the reliability and validity of the constructs used in the study which is discussed below.

8.3.6.1 Reliability

The reliability is the extent to which a variable or a set of variables is consistent in what it intends (Hair et al. 2009). We first computed the Cronbach alpha (α) value for all the constructs which range from 0.74 to 0.88, thus exceeding the recommended value, that is, 0.70 (Hair et al. 2009). The study also identified that the values of composite reliability are within acceptable range, that is, between 0.95 and 0.98, greater than 0.70 benchmark (Malhotra and Dash 2010). All scales are found to be reliable.

8.3.6.2 Validity

The study assessed the convergent and discriminant validity of the measurement scales. Convergent validity is determined using average variance extracted (Malhotra and Dash 2010). As shown in Table 8.5, the average variance extracted values for all constructs is found to be above the threshold criterion of 0.50 (i.e. between 0.62 and 0.75), indicating convergent validity of the constructs. The study further assessed the discriminant validity (i.e. degree to which the construct is distinct from other constructs) of all the measurement scales. Further discriminant validity is also established as the correlation estimates are less than square root of AVE.

8.4   Results of Hypotheses Testing

The study hypotheses, that is, $H1$, $H3$ and $H4$, are tested using structural equation modelling (SEM) (Fig. 8.2). Based on SEM result, the hypothesis $H1$, indicating significant and positive impact of market orientation on marketing capability, is accepted (CR = 8.6 and SRW = 0.93). The results also confirmed positive relationship of marketing capability with competitive advantage (CR = 3.69 and SRW = 0.32) and business performance (CR = 3.62 and SRW = 0.25). Hence, hypotheses $H2$ and $H4$ stand accepted.

8.4.1   Mediation Results

To assess the indirect role of mediating variable, the study framed three contrasting models—fully (i.e. the model with indirect relationship between independent and outcome variables, with paths from independent and mediating variables as well as from mediating to outcome variables), partially mediated model (i.e. the model with the addition of a direct path from independent variable to outcome variable) and non-mediating model (a direct relationship between independent variable to outcome variable, with no path from mediating variable to outcome variable) (Arnold et al. 2007). The chi-square difference test was used for selection among the models. The mediating relationships results are given in Table 8.6.

   While estimating the effects of marketing capability on market orientation and competitive advantage linkage, the study identifies significant difference among the three models implying that model with better fitness indices is to be selected. The partially mediating model offered better-fit indices than fully mediating model. However, since SRW for two relationships, that is, market orientation–competitive advantage and marketing capability–competitive advantage, were above one, multi-collinearity issue is present, which could be verified further in future analysis. The result confirms marketing capability as a partial mediator in market orientation

Table 8.5 Composite reliability, average variance extracted, correlation matrix and Cronbach's alpha

Constructs	Composite reliability	Average variance extracted	Correlation				Cronbach's alpha
			Marketing capability	Market orientation	Environmental turbulence	Competitive advantage	
Marketing capability	0.98	0.72	0.85[a]	–	–	–	0.88
Market orientation	0.95	0.62	0.73	0.83[a]	–	–	0.83
Environmental turbulence	0.96	0.73	0.53	0.57	0.85[a]	–	0.74
Competitive advantage	0.98	0.75	0.29	0.15	0.06	0.87[a]	0.88

[a]Values in the diagonal of correlation matrix are the square root of AVE

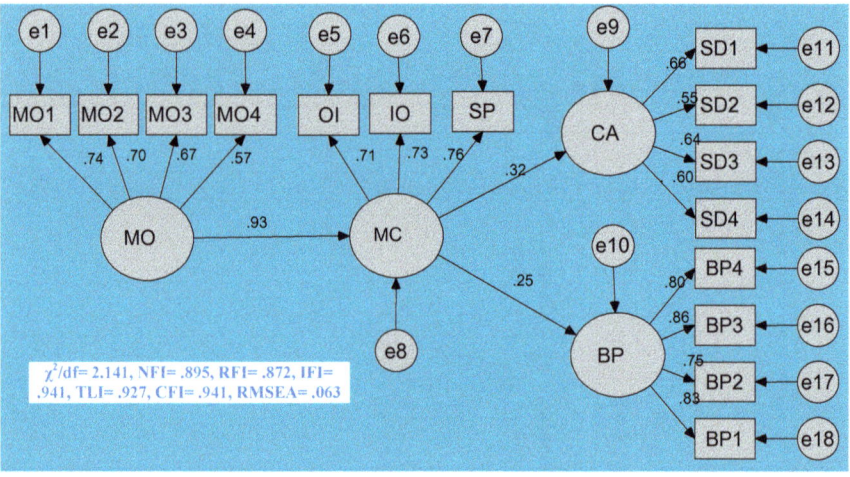

Fig. 8.2 Market orientation, marketing capability, competitive advantage and business performance relationship (*Note* MO = Market orientation, MC = Marketing capability, CA = Competitive advantage, BP = Business performance, MO1 = Customer need-focused intelligence generation, MO2 = Customer satisfaction-focused intelligence, MO3 = Intelligence dissemination, M4 = Responsiveness, OI = Outside-in, IO = Inside-out, SP = Spanning, SD1 = Online bank services, SD2 = Cash/fund processing time, SD3 = ATMs' service quality, SD4 = Draft/cheque processing, BP1 = ROA, BP2 = ROI, BP3 = Market share growth, BP4 = Profitability, e1–e18 = errors variances for model items)

and CA relationship. Hence, *H*2a is partially accepted. Further, mediating role of marketing capability in market orientation and business performance is also accepted. Among the three models, non-mediating model fails to be accepted as its RMSEA has come out to be above 0.08. Since the study found insignificant difference between fully and partially mediating models, both the models are accepted. The use of Sobel test for fully mediating model suggested indirect and significant (critical ratio = 2.14 and *p* = 0.01) impact of marketing capability on market orientation and business performance. On the other hand, partial mediation effect is examined through SEM and Sobel test. The SEM result indicated insignificant relationship (critical ratio = −0.62 and *p* = 0.54) between market orientation and business performance. Similarly, Sobel test also revealed insignificant relationship between market orientation, marketing capability and business performance (i.e. Sobel *t* = 0.82 and *p* = 0.42). Hence, *H*2b is accepted. The study finds significant difference among full, partial and non-mediating models while testing the influence of competitive advantage on marketing capability and business performance relationship. Among the above three models, partially mediating model was accepted. The value of critical ratio (i.e. 2.48) advocated direct and significant relationship between marketing capability and business performance relationship. On the other hand, the use of Sobel test recommended insignificant relationship between marketing capability, competitive advantage and business performance. Thus, the hypothesis *H*5 is partially accepted.

Table 8.6 Mediation models

Models	χ^2/df	Normed fit index	Relative fit index	Incremental fit index	Tucker–Lewis index	Comparative fit index	Root mean square of error approximation
MO–MC–CA							
Fully mediating	2.74	0.89	0.86	0.93	0.90	0.93	0.0 8
Partially mediating	2.58	0.89	0.86	0.94	0.91	0.93	0.07
Non-mediating	3.22	0.88	0.83	0.92	0.88	0.92	0.09
MO–MC–BP							
Fully mediating	2.07	0.94	0.92	0.97	0.96	0.97	0.06
Partially mediating	2.11	0.94	0.92	0.97	0.96	0.97	0.06
Non-mediating	3.09	0.94	0.91	0.96	0.94	0.96	0.08
MC–CA–BP							
Fully mediating	2.49	0.91	0.88	0.94	0.93	0.94	0.07
Partially mediating	2.36	0.92	0.89	0.95	0.93	0.95	0.07
Non-mediating	2.80	0.96	0.93	0.97	0.96	0.97	0.079

8.4.2 Moderation Results

The study used the procedure given by Zhao and Cavusgil (2006) for examining the moderation effect of environmental turbulence using SPSS 17 and AMOS 20. Two models—unconstrained (where all paths are allowed to move freely) and constrained (where paths are constrained fixed to be equal) models—are developed. Based on moderate and high environmental turbulence groups, unconstrained model was developed to identify the presence of insignificant paths for deletion in the two groups. One insignificant path, that is, between competitive advantage and business performance, was identified and deleted. The chi-square value and degree of freedom of both models were compared using Excel. Since, there does not exist any difference in the models, H6a and H6b are not accepted.

8.5 Discussions

The primary aim of the study is to analyse the impact of market orientation as an antecedent to marketing capability vis-à-vis competitive advantage and business performance in the banking sector. In this context, the study found significant and positive linkage of market orientation and marketing capability. These results support the findings of other studies such as Vorhies and Harker (2000), Hooley et al. (2005), and Ngo and O'Cass (2012) which remarked that market-oriented organisation is more capable in building marketing capability. Further, marketing capability developed through such activities enhances an organisation's ability to understand what customers' expect from marketplace offerings and what is to be delivered to them in the market (Ngo and O'Cass 2012). Moreover, the result established that marketing capability is a mediating factor in market orientation and competitive advantage vis-à-vis market orientation and business performance linkages. In line with these findings, scholars, namely Murray et al. (2011) and Ngo and O'Cass (2012), remarked that market orientation activities aid an organisation in building marketing capability by using market-based knowledge and information. Further, the results also reveal partial mediating role of competitive advantage in marketing capability and business performance relationship. Further, the study does not find any moderating role of environmental turbulence in marketing capability, competitive advantage and business performance. This might be because of the fact that commercial Indian banks equipped with marketing capability are proactive in predicting market changes. For instance, with the help of outside-in capabilities, commercial banks are able to predict changes in the customers' needs and requirements, in advance, as banks have maintained long-term and strong relationship with their customers. Further, through spanning capabilities, the banks are capable of changing the marketing practices accordingly. Similarly, Web technology capabilities (inside-out capabilities) aid retail banks in identifying turbulence in technological environment.

8.6 Managerial Implications

The study makes number of academic and managerial contributions. The study advances an in-depth understanding of the significance of marketing capability in attaining business performance in the following manner. First, it offers new insight into marketing theory by identifying four perspectives of marketing capability, from the literature, which included operational, marketing mix, intellectual capital and competition. This study discussed marketing capability from a broader aspect that is operational, which classified marketing capability as outside-in, inside-out and spanning. These capabilities empower managers with competitive skills and knowledge to capture information about the customers' requirements (both internal and external) and the competitors. Second, the study contributes to marketing management literature by highlighting on major dimensions of market orientation (i.e. customer need-focused intelligence generation, customer satisfaction-focused intelligence generation, intelligence dissemination and responsiveness) that is required to develop marketing capability in an organisation to achieve competitive advantage and business performance. Hence, market orientation functions as the marketing support capability that expedites the development of marketing capability. For instance, market orientation activities (such as customer need-focused intelligence generation and customer satisfaction-focused intelligence generation) undertaken by bank help it in providing more accurate and relevant information about customers' unfulfilled needs and their post-purchase satisfaction level. This information facilitates the bank's manager in understanding customers' problems more precisely and guides him/her in taking necessary steps to overcome these problems. Hence, all information generation activities can pave ways for comprehending customers and their needs which represents outside-in capabilities of an organisation. Further, dissemination of such intelligence among the concerned employees leads to the development of inside-out capabilities of an organisation. For example, if the customers express unhappiness regarding the behaviour of the employees, then the bank either through constant guidance or training programmes can motivate its employees to improve their skills and abilities. Furthermore, responsiveness to the intelligence generated allows an organisation to build spanning capabilities such as new product development, pricing capabilities. Further, marketing capability developed and nurtured through learning (market knowledge about customer needs) and experimentation (past experience in forecasting and responding to customer needs) can enable an organisation to generate several benefits over its competitors such as increased customer base and enhanced performance in terms of net growth, profitability, ROI, ROA, etc.

8.7   Limitations and Directions for Future Research

The researchers come across with certain unavoidable limitations while conducting the study. The study was conducted from operational based perspective—outside-in, inside-out and spanning. Thus, comprehending marketing capability based on other perspectives, that is, intellectual capital, marketing mix, and competition and identifying which perspective is more comprehensive in grasping the concept of marketing capability and attaining business performance could be an important basis for research. Further, since the study has focused only on marketing capability, the other organisational capabilities such as technology, R&D, operations, financial may have a stronger impact on business performance. Thus, which capabilities that is, either marketing or other organisational, contribute more towards achieving business performance could also prove to be an interesting line for future research. As the study is cross-sectional in nature, a longitudinal study is suggested for understanding and validating the various relationships identified by the study. The study stresses on the significant role of market orientation in the development of marketing capability, and other variables such as culture, innovation, marketing employee development capabilities and organisation's efficiency on marketing capability–business performance relationship were excluded which could be examined for further development of marketing capability concept.

References

Arnold, K. A., Turner, N., Barling, J., Kelloway, E. K., & McKee, M. C. (2007). Transformational leadership and psychological well-being: The mediating role of meaningful work. *Journal of Occupational Health Psychology, 12*(3), 193–203.

Azizi, S., Movahed, S. A., & Khah, M. H. (2009). The effect of marketing strategy and marketing capabilities on business performance. Case study: Iran's medical equipment sector. *Journal of Medical Marketing, 9*(4), 309–317.

Cadogan, J. W., Hooley, G., Douglas, S. P., Matear, S., & Greenley, G. (2002). Measuring marketing capabilities: A cross-national study. In *Joint ANZMAC/EMAC Symposium Marketing Networks in a Global Marketplace*. Perth, Australia, December 1–8.

Chahal, H., & Kaur, J. (2014). Development of marketing capabilities scale in banking sector. *Measuring Business Excellence, 18*(4), 65–85.

Day, G. S. (1994). The capabilities of market-driven organizations. *Journal of Marketing, 58*(4), 37–52.

Day, G. S. (2011). Closing the marketing capabilities gap. *Journal of Marketing, 75*(4), 183–195.

Fahy, J., Hooley, G., Cox, T., Beracs, J., Fonfara, K., & Snoj, B. (2000). The development and impact of marketing capabilities in Central Europe. *Journal of International Business Studies, 31*(1), 63–81.

Greenley, G. E., Hooley, G. J., & Rudd, J. M. (2005). Market orientation in a multiple stakeholder orientation context: Implications of marketing capabilities and assets. *Journal of Business Research, 58*(11), 1483–1494.

Guenzi, P., & Troilo, G. (2007). The joint contribution of marketing and sales to the creation of superior customer value. *Journal of Business Research, 60,* 98–107.

Hair, J. F., Black, W. C., Babin, B., Anderson, R. E., & Tatham, R. L. (2009). *Multivariate Data Analysis* (7th ed.). Uppersaddle River NJ: Prentice Hall.

Hooley, G., Fahy, J., Cox, T., Beracs, J., Fonfara, K., & Snoj, B. (1999). Marketing capabilities and firm performance: A hierarchical model. *Journal of Market Focused Management, 4*(3), 259–278.

Hooley, G. J., Greenley, G. E., Cadogan, J. W., & Fahy, J. (2005). The performance impact of marketing resources. *Journal of Business Research, 58*(1), 18–27.

Jaworski, J. B., & Kohli, A. K. (1993). Market Orientation: Antecedents and consequences. *Journal of Marketing, 57,* 53–70.

Kaleka, A., & Morgan, N. A. (2017). How marketing capabilities and current performance drive strategic intentions in international markets. *Industrial Marketing Management.* https://doi.org/10.1016/j.indmarman.2017.02.001.

Krasnikov, A., & Jayachandran, S. (2008). The relative impact of marketing research-and-development and operations capabilities on firm performance. *Journal of Marketing, 72*(4), 1–11.

Lafferty, B. A., & Hult, G. T. M. (2001). A synthesis of contemporary market orientation perspectives. *European Journal of Marketing, 35*(1/2), 92–109.

Lindblom, A. T., Olkkonen, R. M., Mitronen, L., & Kajalo, S. (2008). Market-sensing capability and business performance of retail entrepreneurs. *Contemporary Management Research, 4*(3), 219–236.

Malhotra, N. K., & Dash, S. (2010). *Marketing research: An applied orientation* (6th ed.). New Delhi ND: Pearson Education.

Meade, A. W., Watson, A. M., & Kroustalis, C. M. (2007). *Assessing common methods bias in organizational research.* Paper presented at the 22nd Annual Meeting of the Society for Industrial and Organizational Psychology, New York.

Merrilees, B., Rundle-Thiele, S., & Lye, A. (2011). Marketing capabilities: Antecedents and implications for B2B SME performance. *Industrial Marketing Management, 40,* 368–375.

Moller, K., & Antilla, M. (1987). Marketing capability—A key success factor in small business? *Journal of Marketing Management, 3*(2), 185–220.

Morgan, N. A., Vorhies, D. W., & Mason, C. H. (2009). Market orientation marketing capabilities and firm performance. *Strategic Management Journal, 30*(8), 909–920.

Murray, J. Y., Gao, G. Y., & Kotabe, M. (2011). Market orientation and performance of export ventures: The process through marketing capabilities and competitive advantages. *Journal of the Academy of Marketing Science, 39*(2), 252–269.

Najafi Tavani, S., Sharifi, H., & Najafi Tavani, Z. (2016). Market orientation, marketing capability, and new product performance: The moderating role of absorptive capacity. *Journal of Business Research, 69*(11), 5059–5064.

Narver, J. C., & Slater, S. F. (1990). The effect of a market orientation on business profitability. *Journal of Marketing, 54,* 20–35.

Nath, P., Nachiappan, S., & Ramanathan, R. (2010). The impact of marketing capability operations capability and diversification strategy on performance: A resource-based view. *Industrial Marketing Management, 39*(2), 317–329.

Ngo, V. L., & O'Cass, A. (2012). Performance implication of market orientation marketing resources and marketing capabilities. *Journal of Marketing Management, 28*(1/2), 173–187.

Pham, T. S. H., Monkhouse, L. L., & Barnes, B. (2017). Influence of relational capability and marketing capabilities on the export performance of emerging market firms. *International Marketing Review, 34*(5), 606–628.

Podsakoff, P. M., MacKenzie, S. B., Podsakoff, N. P., & Lee, J. Y. (2003). Common method biases in behavioural research: A critical review of the literature and recommend remedies. *Journal of Applied Psychology, 88*(5), 879–903.

Qureshi, S., & Mian, S. A. (2010). Antecedents and outcomes of entrepreneurial firms marketing capabilities: An empirical investigation of small technology based firms. *Journal of Strategic Innovation and Sustainability, 6*(4), 28–45.

Ros, S. C., Cruz, T. F., & Cabanero, C. P. (2010). Marketing capabilities stakeholders' satisfaction and performance. *Service Business: An International Journal, 4*(3/4), 209–223.

Tsai, M. T., & Shih, C. M. (2004). The impact of marketing knowledge among managers on marketing capabilities and business performance. *International Journal of management, 21*(4), 524–530.

Vijande, L. S., Perez, M. J. S., Gutierrez, J. A. T., & Rodriguez, N. G. (2012). Marketing capabilities development in small and medium enterprises: Implications for performance. *Journal of Centrum Cathedra, 5*(1), 24–42.

Vorhies, D. W. (1998). An investigation of the factors leading to the development of marketing capabilities and organizational effectiveness. *Journal of Strategic Marketing, 6*(1), 3–24.

Vorhies, D. W., & Harker, M. (2000). The capabilities and performance advantages of market-driven firms: An empirical investigation. *Australian Journal of Management, 25*(2), 145–171.

Vorhies, D. W., Harker, M., & Rao, C. P. (1999). The capabilities and performance advantages of market-driven firms European. *Journal of Marketing, 33*(11/12), 1171–1202.

Vorhies, D. W., & Morgan, N. A. (2005). Benchmarking marketing capabilities for sustainable competitive advantage. *Journal of Marketing, 69*(1), 80–94.

Vorhies, D. W., & Yarbrough, L. (1998). Marketing's role in the development of competitive advantage: Evidence from the motor carrier industry. *Journal of Market Focused Management, 2*(4), 361–386.

Weerawardena, J. (2003). The role of marketing capability in innovation-based competitive strategy. *Journal of Strategic Marketing, 11*(1), 15–35.

Zhao, Y., & Cavusgil, S. T. (2006). The effect of supplier's market orientation on manufacturer's trust. *Journal of Industrial Marketing Management, 35,* 405–414.

Chapter 9
Examining the Impact of Cultural Intelligence on Knowledge Sharing: Role of Moderating and Mediating Variables

Jeevan Jyoti, Vijay Pereira and Sumeet Kour

Abstract Globalisation of world has brought lot of challenges for individuals and organisations in the form of cultural diversity management. In this perspective, cultural intelligence is an ability, which can enhance an employee's skill to communicate with individuals belonging to his/her culture as well as host region nationals. The study aims at analysing the moderating role played by work experience between cultural intelligence (CQ) and cross-cultural adjustment (CCA) relationship. Further, the mediating role is played by cross-cultural adjustment between cultural intelligence and knowledge sharing relationship. 530 bank managers working in nationalised banks operating in Delhi (North India) have been contacted for the study. In order to establish normality of the data, 18 respondents have been deleted by inspecting boxplots. Therefore, the effective sample came to 512. Confirmatory factor analysis (CFA) has been used to validate the scale, and to check the hypotheses, structural equation modelling (SEM) has been used. The result reveals that work experience moderates between CQ and CCA. The findings further reveal that CCA mediate between CQ and knowledge sharing relationship. The study is cross-sectional in nature. Further, the role of only one moderating variable, i.e. work experience, has been explored between CQ and CCA relationship. The study contributes towards cultural intelligence theory. Cultural intelligence acts as an essential tool in selection of managers who can work effectively in cross-cultural context. Culturally intelligent managers are talented and interactive which helps them to give their best performance. These managers can be sent for overseas assignments as they are able to communicate successfully with individuals belonging to dissimilar cultural backgrounds.

J. Jyoti · S. Kour (✉)
Department of Commerce, University of Jammu, Jammu, Jammu and Kashmir, India
e-mail: sumeetask@gmail.com

J. Jyoti
e-mail: jyotigupta64@rediffmail.com

V. Pereira
University of Wollongong, Dubai Campus, Dubai, United Arab Emirates
e-mail: VijayPereira@uowdubai.ac.ae

© Springer Nature Singapore Pte Ltd. 2019
H. Chahal et al. (eds.), *Understanding the Role of Business Analytics*,
https://doi.org/10.1007/978-981-13-1334-9_9

Keywords Cross-cultural adjustment · Cultural intelligence · Knowledge sharing and previous work experience · India

9.1 Introduction

With the rising globalisation, there is an increasing need to know how to successfully communicate with individuals belonging to diverse culture. The organisations today require managers, who are aware and have knowledge of diverse cultures as they have to interconnect with individuals from various cultures. Therefore, managers need to have global expertise to become an effective leader. In this perspective, cultural intelligence (CQ) is the capability, which enhances person's skill to communicate with individuals outside their culture and nation. It is the skill and trait that let one to successfully network with new cultural settings (MacNab et al. 2012). Culturally intelligent managers can sense, adjust, reason and act on cultural signals promptly in conditions distinguished by cultural diversity. Lower level of CQ gives rise to cultural encounter and unhealthy associations. More and more organisations are stating the necessity for managers who rapidly adapt to several cultures and work in international teams (Earley and Peterson 2004, p. 100). CQ is emerging notion with narrow investigation on this. Hence, the study generalises the concept of CQ in Indian context (Tsang and Kwan 1999) due to its varied cultures. Further, the present study also assess its effect on knowledge sharing via CCA and the role played by work experience in between CQ and knowledge sharing relationship.

India is a culturally diverse country, which needs managers, who can efficiently manage the diverse workforce. Most of the studies conducted on CQ have focused on the concept (Earley and Peterson 2004; Ng and Earley 2006; Triandis 2006; Turner and Trompenaars 2006; Kumar et al. 2008; Thomas et al. 2008; Crowne 2009; Van Dyne et al. 2010; Blasco et al. 2012). Review of literature indicated that CQ significantly affects CCA (Ang et al. 2007; Kumar et al. 2008; Lee 2010; Lee and Sukoco 2010; Ramalu et al. 2010, 2011; Huff 2013; Malek and Budhwar 2013; Huff et al. 2014; Lee and Kartika 2014), and most of them are done in MNCs and focused mainly on expatriates' CQ. There are few studies about CQ of Indian managers (Jyoti and Kour 2015; Jyoti et al. 2015), who have to work with various cultural workforce, and few studies have been conducted on the relationship between CCA on knowledge sharing (Lee and Kartika 2014), which demands research in this area in India. Lee and Kartika (2014) founded that CCA significantly predicts knowledge sharing. They concluded that expatriates who adapt well to the host region become a vital part of knowledge transfer from parent company to subsidiary company and enhance the organisational performance. Therefore, the study will inspect the effect of CQ on cross-cultural adjustment and further the impact of cross-cultural adjustment on knowledge sharing. Further, various empirical researches have shown the positive affect of work experience on cultural intelligence (Crowne 2008; Moon 2010; Moon et al. 2012; Lee and Kartika 2014) and cross-cultural adjustment (Lee and Kartika 2014; Peltokorpi and Froese 2012; Huff et al. 2014). In contrast, researchers have also found that experience has an insignificant influence on cultural intelligence (MacNab and Worthley 2012; Lee 2010; Gupta et al. 2013). Lee and Sukoco (2010) have revealed the moderating role of experience in between CQ and CCA relationship. Hence, there is dearth of consent concerning this relationship. So, to evidently understand the relationship of work experience with CQ and CCA, the present study will examine the role played by work experience between cultural intelligence and CCA.

Therefore, this study will try to cover all the possible gaps. So, the objective of the study is to analyse the mediating role played by cross-cultural adjustment between cultural intelligence and knowledge sharing relationship. Further, the role of work experience between CQ and CCA shall also be explored.

9.2 Objectives of the Study

1. To examine the impact of cultural intelligence on cross-cultural adjustment.
2. To study the effect of CCA on knowledge sharing.
3. To analyse the mediating role played by cross-cultural adjustment between CQ and knowledge sharing.
4. To study the moderating role played by work experience between CQ and CCA.

9.3 Review of Literature and Hypotheses Development

9.3.1 Cultural Intelligence, Experience and Cross-Cultural Adjustment

Experience is a time element (Jyoti and Kour 2017b; Goodman et al. 2001). In the present study, experience denotes the direct observation in culturally connected actions or the state of being affected by such observation (Takeuchi and Chen 2013). The experience is the events that have occurred in the past and is presently occurring (Goodman et al. 2001). Earlier studies have indicated that cultural intelligence positively affects CCA (Jyoti and Kour 2015, 2017a, b; Wang 2016). Further, researchers found that the effect of cultural intelligence on cross-cultural adjustment is magnified if the previous cross-cultural work experience is positive (Takeuchi et al. 2005; Lee and Sukoco 2010). In this perceptive, managers, having more cross-cultural experience, are more likely to develop comprehensive cognitive schemata (Lee and Sukoco 2010). So, managers with more CQ level as well as higher international work experience adjust and accomplish more successfully in the host region (Lee and Sukoco 2010). Cross-cultural work experience enhances the confidence and exposure to effectively communicate with individuals outside their culture, which aids them to adjust to their host region (Bhaskar-Shrinivas et al. 2005 cited in Moon et al. 2012). Previous international work experience offers an individual with the means of foreseeing what an overseas assignment contains, rises the probability of realistic expectations, decreases uncertainty and thereby simplifying the adjustment (Black et al. 1992). Out of home state, work experience helps culturally intelligent individuals/managers to learn suitable work behaviours and to learn how to communicate with locals of host region (Lee 2010), which aid them to adjust to host region. When the culturally intelligent managers have more experience of working in host region, they have a tendency to adjust more effortlessly during their overseas assignments. Whereas there are some studies, which have revealed an insignificant influence of cross-cultural work experience on cross-cultural adjustment (Hechanova et al. 2003; Puck et al. 2008; Shaffer et al. 1999 cited in Moon et al. 2012). Therefore, there is lack of agreement regarding this relationship. Thus, to clearly understand the role of previous work experience in the relationship between cultural intelligence and cross-cultural adjustment, the present study will analyse the integrative model, wherein positive work experience moderates the impact of cultural intelligence on cross-cultural adjustment. Therefore, previous work experience magnifies/strengthens the CQ and CCA relationship.

Hypothesis 1 Work experience moderates the relationship between cultural intelligence and cross-cultural adjustment.

9.3.2 Cultural Intelligence, Cross-Cultural Adjustment and Knowledge Sharing

The concept of cultural intelligence (CQ) has been developed by Earley and Ang (2003). CQ is the capability of the individuals to function successfully in multi-cultural settings (Ang and Van Dyne 2008). Individuals with CQ level have the capability to handle unclear and confusing circumstances. They think deeply about the situation or conditions and make suitable adjustments to how they understand, relate and lead in the context of these various cultures. It is adjustable state that can be developed over a period of time. Cultural intelligence is a multi-dimensional concept containing cognitive, meta-cognitive, behavioural and motivational dimensions (Ang et al. 2007). This phenomenon positively affects CCA (Jyoti and Kour 2015, 2017a, b; Wang 2016; Mehra and Tung 2017). Cross-cultural adjustment is the psychological ease an individual has in the host region (Black and Stephens 1989; Gregersen and Black 1990). Researchers in the cross-cultural studies have revealed the positive impact of CQ on CCA (Ramalu et al. 2010, 2011; Lee and Sukoco 2010). CQ has a direct impact on the CCA as it aids the individuals to adapt more effortlessly to the host region environment (Earley and Ang 2003). The two dimensions of CQ, i.e. meta-cognitive and motivational dimension, are positively associated with all the three dimensions of cross-cultural adjustment, i.e. general, work and interaction adjustment (Ramalu et al. 2010) as it accelerates the cultural learning process and develops essential interest in various other cultures. Further, behavioural cultural intelligence is positively associated with CCA as it has the ability to differ behaviours, which help to adapt in an unaware and unfamiliar environment (Ramalu et al. 2011; Kumar et al. 2008). Culturally intelligent managers have flexible behaviour, which aids them to adapt in multi-cultural setting. Cognitive cultural intelligence is the knowledge element of cultural intelligence, which positively correlates to all the dimensions of cross-cultural adjustment (Ramalu et al. 2011; Kumar et al. 2008). Individuals who have higher level of CQ are more able to adapt in the host region environment.

Cross-cultural adjustment is the process by which an expatriate feel psychologically comfortable in the new cultural environment and harmonises with it (Huang et al. 2005). CCA is multi-dimensional concept comprising general adjustment, work adjustment and interaction adjustment. It is one of the primary outcomes in an expatriate assignment that affect the development of secondary or more distal expatriate adjustment such as stress (Hechanova et al. 2003), organisational commitment (Naumann 1993; Shaffer and Harrison 1998), job satisfaction (Takeuchi et al. 2002), performance (Kim and Slocum 2008; Shay and Baack 2006) and turnover intentions (Hechanova et al. 2003). This phenomenon positively affects knowledge sharing (Lee and Kartika 2014). Knowledge is a critical asset for organisations (Nonaka 1994). Knowledge sharing relates to exchanging events, experience, thought or understanding of anything with a hope to increase more understanding and insight understanding about something for temporary curiosity (Sohail and Daud 2009). Cultural diversity within the workplace could impact knowledge sharing processes. Studies have revealed that culture plays a vital role in KS processes (De Long 1997; Gold et al. 2001; Kayworth and Leidner 2003). Culture has been seen as a barrier in knowledge sharing process (De Long and Fahey 2000; McDermott and O'Dell 2001). Culture is the set of characteristics a particular set of people have with respect to language, social habits, religion, habits, cuisine, art and music (Jyoti and Kour 2015). Knowledge sharing is influenced by the prevailing culture (Peng et al. 2008). Therefore, when the expatriate adjust themselves to the cultural difference, they are able to share their knowledge to the host nationals in a better manner. Lee and Kartika (2014) revealed that adjustment of the expatriate has a positive impact on expatriate's knowledge transfer. Expatriates are the source of knowledge transfers from parent company to the foreign subsidiary or from the host county knowledge to back to the parent company (Lee and Kartika 2014, p. 5486). Expatriate who adapt successfully in the new and unfamiliar environment are more able to complete their assignment and create an understanding about corporate challenges, and it became a significant part of the knowledge transfer from parent company to the subsidiary or from the host country back to the parent company (Lee and Kartika 2014, p. 5490). Expatriate who adjust themselves in the host region have high knowledge, greater ability to communicate and share knowledge with colleagues (Paik and Shon 2004). Expatriates are the home region assignee that holds key positions in host regions and transfers or share their knowledge with colleagues at workplace (Harzing 2001). Well-adjusted expatriates do not have difficulty in creating and sharing knowledge as they interact effectively with host nationals and interaction is core in contributing and developing new knowledge (Nonaka 1994). Managers who adjust themselves in host region are able to share new and innovative ideas, which in turn helps organisation to increase its performance. Well-adjusted managers act as source of knowledge transfer from home region to host region (Lee and Kartika 2014). The manager who is well adjusted and is prepared to face the challenge of new culture and unfamiliar concepts will not hesitate to share and implement new idea at the workplace. Therefore, it can be concluded from above discussion that cross-cultural adjustment significantly affects knowledge sharing.

CQ significantly affects CCA as it comforts the individuals to adjust more effortlessly to the host region (Earley and Ang 2003). Indian managers have to communicate with people belonging to diverse cultural background and make several adjustments as each region (western, eastern, northern and southern) of India have diverse values, languages and belief system (Banerjee 2013). Further, expatriate is vital part of the knowledge transfer from parent company's knowledge to the foreign subsidiary or from host country knowledge to be transferred back to the parent company (Gong 2003). When culturally intelligent managers adjust themselves in host region environment, he/she can easily share their knowledge at the host region and can systematically introduce the new ideas. Culturally intelligent managers who adjust themselves with the general, interaction and work environment of the host region can easily share their knowledge with others as they can effectively communicate with locals and colleagues. When expatriate adopt out of home state assignment, he/she have to make various kinds of adjustment related to general environment (food, clothing, housing facilities, cost of living, etc.), working environment (responsibilities, supervision, working norms, etc.) and have to make interactions with the local nationals and colleagues, and when expatriate adjust themselves in host region environment, he/she is in a better position to share their ideas, views and knowledge. Therefore, to conclude, managers who are culturally intelligent adapt themselves in multi-cultural settings are in the better position to share their knowledge.

Hypothesis 2 Cross-cultural adjustment mediates between cultural intelligence and knowledge sharing relationship (Fig. 9.1).

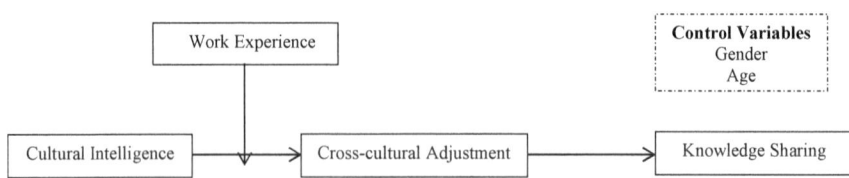

Fig. 9.1 Conceptual framework

9.4 Research Methodology

To make the study objective, the following steps have been taken:

9.4.1 Data Collection

The study population consisted of 530 bank managers working in nationalised banks operating in Delhi (North India). They have been contacted on the basis of random sampling (chit method). There are 2539 nationalised banks operating in Delhi, out of which 10% have been selected with the help of random number table. From each selected bank, two managers (on the basis of hierarchy) have been contacted personally for data generation. All branch managers and immediate junior managers have been contacted for data collection, but in some banks, there were only one manager, in that case, one extra branch has been contacted. Therefore, total 265 banks have been contacted. Structured questionnaire has been used to gather the data. In order to establish normality of the data, 18 respondents have been deleted by inspecting boxplots (Hair et al. 2010). The retained data exhibited normal distributed (skewness = 0.066; Kurtosis = −0.101) which are within the range. Therefore, the effective sample came to 512.

The sample included 286 (56%) male, and majority of the managers (88%) are married. About 29% managers are in the age group 35–40 years followed by 31–34 years (17%). Majority of managers (43%) have 6–10 years of experience of working outside their home state. About 55% of managers can speak 4–5 languages. Majority of managers have 6–10 years of total work experience (24%).

9.4.2 Measures

For the sake of consistency, five-point Likert scale was used for measuring the variables ranging from strongly disagree (1) to strongly agree (5).

Cultural Intelligence 20-items CQ scale developed and validated by Ang et al. (2007), having four factors, have been used. The inventory comprises four items for meta-cognitive CQ, five for motivational CQ, six for cognitive CQ and five for behavioural CQ. Sample items include "Know the legal and economic systems of other cultures", and "Enjoy interacting with people from different cultures".

Cross-Cultural Adjustment Black and Porter (1991) 14-items scale has been used to measure CCA. The scale comprises seven items of general adjustment, three items of work adjustment and four items of interaction items. Sample items include "Adjust myself to interact with host people on day to day basis in state I am posted other than home" and "Adjust myself to the performance standards and expectations at work in state I am posted other than home".

Knowledge Sharing Ten items have been used to measure KS (Ramayah et al. 2014). Sample items contain "I express ideas and thoughts in meetings" and "I propose problem-solving suggestions in team meetings".

9.5 Results

9.5.1 Exploratory Factor Analysis (EFA)

EFA has been conducted to recognise the dimensions of various scales used in the study. Varimax rotation of principle component analysis has been used. To test the appropriateness of a factor analysis KMO measure of sampling adequacy has used, wherein values greater than 0.50 are acceptable (Hair et al. 2010), indicating its significance for further analysis. The statement with factor loading less than 0.50 has been removed (Hair et al. 2010). The KS scale contains ten items that got reduced to five items and joined under one factor. Likewise, cultural intelligence scale originally contains 20 items that got reduced to 14 items and converged under four factors (viz. meta-cognitive, cognitive and motivational). Lastly, the CCA scale contains 14 items, which has been reduced to ten items and joined under the three factors (viz. general adjustment, work adjustment and interaction adjustment). For all the constructs, KMO value is greater than 0.78 and total variance explained for all the constructs is above eighty per cent (Hair et al. 2010). Detailed results are presented in Table 9.1.

Table 9.1 Results of exploratory factor analysis

Factor	M	SD	FL	C	E.V.	V.E. (%)	KMO	Cronbach alpha
Cultural intelligence	4.11	0.71				85.483	0.887	0.934
Meta-cognitive	4.18	0.89			3.547	25.337		0.871
MOG1	4.09	0.87	0.774	0.812				
MOG2	4.19	0.90	0.807	0.806				
MOG3	4.27	0.91	0.764	0.781				
Cognitive	4.01	1.00			3.515	25.107		0.894
COG3	4.05	1.11	0.870	0.887				
COG4	3.97	1.15	0.867	0.864				
COG5	3.91	1.17	0.805	0.737				
Motivation	4.11	0.93			2.573	18.380		0.955
MOT1	4.14	1.00	0.930	0.956				
MOT2	4.07	1.00	0.928	0.953				
MOT3	4.14	0.90	0.914	0.911				
MOT5	4.11	0.93	0.745	0.730				

(continued)

Table 9.1 (continued)

Factor	M	SD	FL	C	E.V.	V.E. (%)	KMO	Cronbach alpha
Behavioural	4.17	0.82			2.332	16.660		0.954
BEH1	4.19	0.85	0.890	0.947				
BEH2	4.18	0.88	0.829	0.838				
BEH3	4.19	0.85	0.889	0.947				
BEH4	4.10	0.92	0.824	0.797				
Cross-cultural adjustment	4.14	0.69				75.033	0.923	0.919
General adjustment	4.15	0.72			3.586	35.858		0.786
GA1	4.24	0.82	0.843	0.772				
GA2	4.33	0.78	0.829	0.754				
GA5	3.93	1.10	0.844	0.790				
GA6	4.10	0.94	0.695	0.659				
Interaction adjustment	4.07	0.85			2.144	21.441		0.759
IA1	4.04	1.12	0.891	0.889				
IA2	4.18	0.83	0.651	0.752				
IA4	4.21	0.82	0.694	0.664				
Work adjustment	4.21	0.70			1.773	17.733		0.879
WA1	4.26	0.80	0.688	0.679				
WA2	4.17	0.81	0.566	0.726				
WA3	4.21	0.75	0.677	0.809				
Knowledge sharing	4.24	0.65			3.175	63.491	0.780	0.850
KS1	4.26	0.70	0.855	0.730				
KS2	4.35	0.77	0.778	0.606				
KS5	4.11	0.98	0.780	0.609				
KS6	4.20	0.88	0.800	0.640				
KS9	4.30	0.75	0.768	0.590				

Key M = mean, SD = standard deviation, FL = factor loading, *C* = communality, E.V. = eigen value, V.
E. = variance explained and KMO = Kaiser-Meyer-Olkin measure of sampling adequacy

9.5.2 *Confirmatory Factor Analysis (CFA)*

To test the validity and reliability of the constructs, CFA has been used. As multiple factors have been emerged after exploratory factor analysis, therefore, second-order factor models have been designed for all the scales. Fit indices of all the models are within the recommended limit as all the values of the absolute goodness of fit (GFI and AGFI), incremental fit (NFI and CFI) and badness of fit (RMSEA and RMR) were within the threshold limit (Table 9.2). Further, convergent validity has been established as all the standardised estimates are above 0.50 and the variance explained by each construct is also above 0.50 (Hair et al. 2010, Table 9.2). In order to test the internal consistency, composite reliability and Cronbach's alpha have been calculated as it is the display the reliability of the construct (Hair et al. 2010). The results demonstrated that alpha values for all constructs are above 0.70 (Table 9.2) and composite reliability for all constructs is also greater than 0.80

Table 9.2 Reliability and validity analysis and fit indices

Scales	Standardised regression weight	Average variance extracted	Composite reliability	Cronbach's alpha	Fit indices
Cultural intelligence		0.93	0.98	0.93	$\chi^2/df = 3.387$ RMR = 0.052 GFI = 0.937 NFI = 0.976 AGFI = 0.910 CFI = 0.983 RMSEA = 0.068
Meta-cognitive CQ	0.87	0.96	0.98	0.87	
Cognitive CQ	0.67	0.96	0.98	0.89	
Motivational CQ	0.60	0.98	0.99	0.95	
Behavioural CQ	0.77	0.98	0.99	0.95	
Cross-cultural adjustment		0.97	0.99	0.91	$\chi^2/df = 4.950$ RMR = 0.030 GFI = 0.948 NFI = 0.952 AGFI = 0.907 CFI = 0.962 RMSEA = 0.088
General adjustment	0.98	0.93	0.98	0.78	
Interaction adjustment	0.91	0.93	0.97	0.76	
Work adjustment	0.98	0.98	0.99	0.88	
Knowledge Sharing		0.94	0.98	0.85	$\chi^2/df = 2.202$ RMR = 0.023 GFI = 0.948 NFI = 0.994 AGFI = 0.903 CFI = 0.928 RMSEA = 0.073
KS1	0.94				
KS2	0.80				
KS5	0.53				
KS6	0.58				
KS9	0.61				

(Table 9.2). Therefore, the composite reliability and Cronbach's alpha values show that the scales are reliable. Furthermore, discriminant validity has also been proved as average variance extracted (AVE) for all the scales is greater than the squared correlation (Fornell and Larcker 1981, Table 9.3).

Table 9.3 Discriminant validity and correlation analysis

Constructs	Cross-cultural adjustment	Knowledge sharing	Cultural intelligence
Cross-cultural adjustment	0.97		
Knowledge sharing	(0.11) 0.34**	0.94	
Cultural intelligence	(0.73) 0.86**	(0.10) 0.32**	0.93

Note Values on the diagonal axis represents the average variance extracted. Values below the diagonal axis are correlation and values in the parentheses represent the squared correlation.
**$p < 0.01$

9.5.3 Common Method Bias

Single source data have been gathered for the present study, which may inflate the relationship. Therefore, common method bias has been checked through common latent factor method (Podsakoff et al. 2003). The results indicated that there is no item, whose difference is greater than 0.20 (Gaskin 2012). The chi-square difference test further confirmed that the two models that are with common latent factor and without common latent factor model are different ($\Delta\chi^2 > 270.336$, $p < 0.001$). Thus, in the present study, common method bias is not the problem.

9.5.4 Examining the Impact of Cultural Intelligence on Knowledge Sharing: Role of Moderating and Mediating Variables

9.5.4.1 Moderation

To check the various hypotheses, structural equation modelling has been used (Byrne 2010). In this study, we have work experience (metric) as moderating variables. To test the moderation of work experience, interaction effect has been used (Little et al. 2007, p. 223).

Product indicator approach has been applied to model the moderating effect of previous work experience (metric) (Chin et al. 1996, 2003). For cultural intelligence, we have four manifest variables (meta-cognitive, cognitive, motivational and behavioural) and previous work experience is metric and observed in nature, which lead to four latent interaction variables (Mog*Exp, Cog*Exp, Mot*Exp and Beh*Exp). Further, to check the moderating effect of previous work experience, conditions described by Baron and Kenny (1986) have been satisfied first. The results demonstrated that previous work experience has insignificant effect on cross-cultural adjustment (SRW = 0.02, $p > 0.05$) and the interaction of CQ and work experience is significantly predicting cross-cultural adjustment (SRW = 0. 21, $p < 0.01$, Table 9.4). Thus, it can be concluded that previous work experience moderates between cultural intelligence and cross-cultural adjustment relationship. Hence, hypothesis 9.1 stands accepted.

9.5.4.2 Mediation

To check the various mediations in the study Preacher and Hayes (2004), methodology has been followed. They suggested that to check the mediation effect, significance of indirect effect should be analysed. Thus, the estimation of the indirect effect, with the Sobel test as well as with a bootstrap approach, to obtain confidence intervals (CIs) has been used in the study.

Table 9.4 Structural equation modeling results for moderation

	Model I	Model II	Model III
CQ → CCA	0.88***	0.88***	0.85***
Work experience → CCA		0.02 (ns)	0.19***
Work experience * CQ → CCA			0.21***
R^2	0.77	0.77	0.80
Covariance		.	.
CQ and work experience		0.11*	0.12*
Work experience and CQ * work experience			0.56***
Work experience * CQ and CQ			0.11*

$*p < 0.05$; $***p < 0.001$

Table 9.5 Bootstrapping results for mediation analysis

Hypothesis	Independent → mediator	Mediator → dependent	Indirect effect	LL 95%/UL 95%
CQ → CCA → KS	0.88***	0.20***	0.18**	0.053/0.464

Note $**p < 0.01$, $***p < 0.001$; $N = 1000$ bootstrapping resamples; LL BCA and UL BCA = lower level and upper level of the bias corrected and accelerated confidence interval
Key CQ = cultural intelligence, CCA = cross-cultural adjustment, KS = knowledge sharing

The results founded significant effect of CQ on cross-cultural adjustment ($SRW_a = 0.88$, $p < 0.001$, Table 9.5) and CCA on knowledge sharing (KS) relationship ($SRW_b = 0.20$, $p < 0.001$, Table 9.5). Further, the Sobel statistic is also found to be significant for the indirect effect of cultural intelligence on knowledge sharing through cross-cultural adjustment (Sobel statistic = 2.851, $p < 0.01$). In addition, bootstrapping results also yielded significant indirect effect of cultural intelligence on knowledge sharing through CCA (0.18, $p < 0.01$, Table 9.5). The upper and lower bound values did not contain zero at 95% confidence interval (Table 9.5). Further, the model yielded moderate fit ($\chi^2/df = 6.527$, RMR = 0.030, GFI = 0.924, AGFI = 0.856, NFI = 0.941, CFI = 0.950, RMSEA = 0.104). Moreover, the absence or presence of control variables did not bring any variation in the hypothesised relationships, so these have not been considered during evaluation (Arnold et al. 2007). Hence, hypothesis 9.2 got accepted.

9.5.4.3 Moderated-Mediation Test

Further, integrated model has been tested wherein the strength of the relationship between cultural intelligence on knowledge sharing through cross-cultural adjustment is conditional on the moderator's value, i.e. previous work experience. The moderated mediation is confirmed when the indirect effect of cultural intelligence on knowledge sharing in the presence of moderating variable is significant. The moderated-mediation effect of the interaction of CQ and previous work experience

Table 9.6 Bootstrapped conditional indirect effect of CQ on knowledge sharing through CCA at value of work experience (moderator)

Moderator	Level	Conditional indirect effect	Boot SE	Boot LL 95%	Boot UL 95%
Experience	High	0.510***	0.070	0.388	0.651
	Low	0.200*	0.394	0.050	0.350

Note $*p < 0.05$; $***p < 0.001$; $N = 5000$ bootstrapping resamples; LL BCA and UL BCA = lower level and upper level of the bias corrected and accelerated confidence interval

through cross-cultural adjustment on knowledge sharing is significant for both the groups as the indirect relations are significant (Table 9.6).

9.6 Discussion

The present study depicts the significance of cultural intelligence (CQ) in increasing knowledge sharing. The study has undertaken three issues: (i) the moderating role of previous work experience in between CQ and cross-cultural adjustment relationship, (ii) mediating role of cross-cultural adjustment (CCA) in between CQ and knowledge sharing and (iii) moderated mediation of previous work experience and cross-cultural adjustment between cultural intelligence and knowledge sharing.

The study found that cultural intelligence positively affects cross-cultural adjustment; i.e., culturally intelligent managers are more adjustable. The findings are consistent with the earlier research (Ramalu et al. 2010, 2011; Jyoti and Kour 2015; Jyoti et al. 2015). Culturally intelligent managers are more adjustable with people of host region. They can effectively handle stresses and cultural shocks. Managers make different types of adjustments concerning to various languages, as India being multi-lingual and multi-ethnic country, and they also have to make different adaptations relating to clothing, food, shopping conditions, etc. (Jyoti and Kour 2015). CQ aids managers to adjust with culturally dissimilar situations. Managers, who have the capability to manage with several types of stress connected with cross-cultural communications, are more able to adapt in a new cultural settings/environment.

The study further revealed that cultural intelligence and cross-cultural adjustment relationship gets strengthened when managers have experience of working in host region. The relationship between cultural intelligence and cross-cultural adjustment gets boosted when the managers' have the experience of working outside their home state. Culturally intelligent managers who possess experience of working in host region are more adapt in cross-cultural settings. These managers have the awareness as well as the knowledge from previous experience about the culture spread in host region. This experience increases the confidence level of managers to communicate in culturally diverse situations. Experienced managers have the skill, understanding as well as knowledge of the host region setting (language, culture, religion, beliefs, values, etc.), which boost the impact of cultural intelligence on CCA. They recall and recollect their earlier cross-cultural meets with host region nationals, which aids them in accomplishing understanding about host region and adjust themselves in that atmosphere. Experienced managers have more prospects and opportunities to communicate with host region nationals and have necessary knowledge and skills that aid them to adjust in culturally dissimilar region (Moon et al. 2012). Having more experience of working in host region does not mean that managers/expatriates are more culturally adjustable, unless they also have greater cultural intelligence (Lee and Sukoco 2010). Hence, work experience acts as a catalyst or moderator between cultural intelligence and CCA. The result is in line with the earlier studies (Takeuchi et al. 2005; Lee and Sukoco 2010; Lee 2010). Managers who are culturally intelligent and also have the experience of working in outside their home regions are more adaptable or adjustable in culturally diverse settings. Therefore, higher the work experience, stronger is the relationship between cultural intelligence and cross-cultural adjustment.

The study revealed that CCA mediates between cultural intelligence and knowledge sharing relationship. The results indicated that CQ significantly influences cross-cultural adjustment (Ramalu et al. 2010, 2011; Lee and Sukoco 2010; Jyoti and Kour 2015; Jyoti et al. 2015), which in turn affects knowledge sharing (Lee and Kartika 2014). High level of CQ increases CCA, which in turn leads to knowledge sharing among the managers. The results revealed that culturally intelligent managers have enhanced level of cross-cultural adjustment. Managers with higher cultural intelligence are more able to adjust with individuals that belong to diverse culture as they are flexible and can handle successfully the cultural shocks and stresses (Jyoti and Kour 2015). Culturally intelligent managers adapt successfully in cross-cultural environment, which in turn helps to express and share their ideas and thoughts with employees working in the organisation. Culturally intelligent managers spend more time in personal interactions (discussion over lunch, through telephone, etc.) and professional conversation with others managers and employees working with them as they can adjust to speak with local people and colleagues. They become an important source of knowledge sharing from home region to host region. Well-adjusted managers complete their out of home state assignments successfully as they have an understanding of the host region situation, which helps them to easily share the knowledge with employees working in host region. Well-adjusted managers regularly update themselves with banking rules and regulations and exchange the same with the employees working with them in host region. Culturally intelligent managers are motivated and confident to communicate in unfamiliar environment, which helps them to adjust themselves in culturally diverse settings. It results in sharing of more knowledge among employees working in the bank. Managers, who adjust themselves in diverse culture, have better understanding of host region, which helps them to easily disseminate their knowledge. Further, results revealed that all the dimensions of CCA, i.e., interaction adjustment, general adjustment and work adjustment, mediate between cultural intelligence and knowledge sharing relationship. Culturally intelligent managers are more able to adapt with the general conditions (food, living conditions, housing conditions, entertainment facilities, health facilities, cost of living, etc.), working conditions (job responsibilities, performance standards, supervisory responsibilities, etc.) and can adjust themselves to socialise with host region nationals, which comforts them to share their ideas and experiences in foreign assignments. Hence, it can be concluded that cross-cultural adjustment mediates in CQ and knowledge sharing relationship.

Further, it has been revealed that cross-cultural adjustment mediates the relationship between the interaction of cultural intelligence and work experience on knowledge sharing. Culturally intelligent managers with experience are more effortlessly adapt themselves in culturally diverse settings/situations, which boost their knowledge sharing capability in the host region. The study revealed that managers, who are culturally intelligent, can effortlessly adapt themselves in diverse cultural situations if they have worked previously under the similar settings to that in which they are transferred with their co-workers or local nationals, which helps them to share knowledge in host region.

To conclude, the study explains the impact of CQ on knowledge sharing through cross-cultural adjustment and the moderating role of language proficiency and work experience. The study shows that cross-cultural adjustment mediates the combined effect of cultural intelligence and work experience on knowledge sharing.

9.7 Implications

9.7.1 Theoretical Implications

The study contributes to theoretical development of CQ concept by Earley and Ang (2003). The present study enriches the knowledge about CQ as an effective cross-cultural skill construct by proving a relationship between CQ and knowledge sharing. It enhances to CQ–CCA literature by assessing the role of work experience in between this relationship. Further, it proved the construct validity and reliability of CQ, CCA and knowledge sharing within culturally different countries like India, which enlarged the generalisability of the scales. The present study has also assessed the moderated mediation of variables (previous work experience and CCA) in CQ and knowledge sharing relationship, which enhances to the CQ literature.

9.7.2 Practical Implications

The study has certain practical implications which are helpful to organisations. CQ can be used as key selection tool. Cultural intelligence scale will help to identify people who can give their best performance in foreign/overseas assignments as they can successfully communicate in host region. By this, those managers who achieve well in national contexts but probably be unproductive in cross-cultural communications could be screened out, which will lessen unnecessary expenses arising from failure of international/out of state assignments. Developing culturally intelligent employees will aid organisations to sustainable competitive advantage. Hence, CQ can be used by organisation as criteria for evaluation and service compensation. Organisations can improve organisational commitment by encouraging teamwork and by providing job security to the managers, who are posted outside home state.

Organisations should motivate their managers to share the knowledge at workplace, as if things are properly shared, managers become aware about what is expected out of them and they can thus create a road map keeping in mind the accessibility of resources, its pros and cons and try to achieve better results for the organisation as well as clients and themselves. Organisations should formulate teams, and work should be assigned to them as it promotes culture of sharing and improves response time towards the clients making the delivery of the services on time without any delays. The absence of such a sharing culture results into lack of employees taking interest in management's objective, and they feel isolated, thinking themselves to be an unimportant part, which results into resistance to new ideas. Organisation should adopt a variety of mechanisms like knowledge management and preserve the knowledge which can be later used by other employees even when one moves out, and this practice would develop a legacy for the organisation. Further, organisation can reward employees to encourage knowledge sharing. It is suggested that to encourage knowledge sharing, an ideas database should be created and that employees should be paid for their contributions. Rewards should be announced for outstanding sharing employee and department. Organisation can use intranet service that acts as a social/public platform, information centre and employee communication gateway. This helps employees to pursue and share information about problems which are common, minimising the need for managers to step in.

9.8   Limitations and Future Research

The paper has several limitations, which can be taken care of in the future. Firstly, the present study is cross-sectional in nature; it is suggested that in future, longitudinal study can be done. Secondly, additional consequences of CQ can be taken into account in the future study for clearly understanding the concept. Lastly, other variables like language proficiency, compensation, type of expatriation, etc., can also be explored between CQ and CCA.

References

Ang, S., & Van Dyne, L. (Eds.). (2008). *Handbook on cultural intelligence: Theory, measurement and applications.* Armonk, NY: M.E. Sharpe.

Ang, S., Van Dyne, L., Koh, C., Ng, K. Y., Templer, K. J., Tay, C., et al. (2007). Cultural intelligence: Its measurement and effects on cultural judgment and decision making, cultural adaptation and task performance. *Management and Organization Review, 3*(3), 335–371.

Arnold, K. A., Turner, N., & Barling, J. (2007). Transformational leadership and psychological well-being: The mediating role of meaningful work. *Journal of Occupational Health Psychology, 12*(3), 193–203.

Banerjee, A. (2013). What's Indian culture? A dive into domestic diversity. Available at http://davidlivermore.com/2013/08/16/whats-indian-culture-a-dive-into-domestic-diversity/. Accessed on 20 Apr 2012.

Baron, R. M., & Kenny, D. A. (1986). The moderator-mediator variable distinction in social psychological research: Conceptual, strategic and statistical considerations. *Journal of Personality and Social Psychology, 51*(6), 1173–1182.

Bhaskar-Shrinivas, P., Harrison, D. A., Shaffer, M. A., & Luk, D. M. (2005). Input-based and time based models of international adjustment: Meta-analytic evidence and theoretical extensions. *Academy of Management Journal, 48*(2), 259–281.

Black, J. S., & Porter, L. W. (1991). Managerial behavior and job performance: A successful manager in Los Angeles may not be successful in Hong Kong. *Journal of International Business Studies, 22*(1), 99–114.

Black, J. S., & Stephens, G. K. (1989). The influence of the spouse on American expatriate adjustment. *Journal of Management, 15*(4), 529–544.

Black, J. S., Gregersen, H. B., & Mendenhall, M. E. (1992). *Global assignments: Successfully expatriating and repatriating international managers.* San Francisco: Jossey-Bass Publishers.

Blasco, M., Feldt, L. E., & Jakobsen, M. (2012). If only cultural chameleons could fly too: A critical discussion of the concept of cultural intelligence. *International Journal of Cross Cultural Management, 12*(2), 229–245.

Byrne, B. M. (2010). *Structural equation modeling with AMOS* (2nd ed.). New York: Routledge.

Chin, W.W., Marcolin, B.L. & Newsted, P.R. (1996). A partial least squares latent variable modeling approach for measuring interaction effects: Results from a montecarlo simulation study and voicemail emotion/adoption study. In *17th International Conference on Information Systems,* Cleveland, OH. Available at http://wojtek.zozlak.org/ZlozoneModeleSkalowaniaLiiniowego/ChinEtAl_1996_A%20PLS%20Latent%20Variable%20Modeling%20Approach%20for%20Measuring%20Interaction%20Effects%20Results%20from%20MC%20Simulation%20Study.pdf. Accessed 21 Jan 2014.

Chin, W. W., Marcolin, B. L., & Newsted, P. N. (2003). A partial least squares latent variable modeling approach for measuring interaction effects: Results from a montecarlo simulation study and an electronic-mail emotion/adoption study. *Information Systems Research, 14*(2), 189–217.

Crowne, K. A. (2008). What leads to cultural intelligence? *Business Horizons, 51*(5), 391–399.

Crowne, K. A. (2009). The relationships among social intelligence, emotional intelligence and cultural intelligence. *Organisation and Management Journal, 6*(3), 148–163.

De Long, D. W. (1997). *Building the knowledge-based organisation: How culture drives knowledge behaviors.* Boston: Center for Business Innovation, Ernst & Young LLP.

De Long, D. W., & Fahey, L. (2000). Diagnosing cultural barriers to knowledge management. *Academy of Management Executive, 14*(4), 113–127.

Earley, P. C., & Ang, S. (2003). *Cultural intelligence: Individual interactions across cultures.* Palo Alto, CA: Stanford University Press.

Earley, P. C., & Peterson, R. S. (2004). The elusive cultural chameleon: Cultural intelligence as a new approach to intercultural training for the global manager. *Academy of Management Learning and Education, 3*(1), 100–115.

Fornell, C., & Larcker, D. F. (1981). Evaluating structural equation models with unobservable variables and measurement error. *Journal of Marketing Research, 18*(1), 39–50.

Gaskin, J. (2012). Group differences. Retrieved from stats tools package. Available at http://statwiki.kolobkreations.com. Accessed 20 Aug 2014.

Gold, A. H., Malhotra, A., & Segars, A. H. (2001). Knowledge management: An organizational capabilities perspective. *Journal of Management Information Systems, 18*(1), 185–214.

Gong, Y. (2003). Toward a dynamic process model of staffing composition and subsidiary outcomes in multinational enterprises. *Journal of Management, 29*(2), 259–280.

Goodman, P. S., Lawrence, B. S., Ancona, D. G., & Tushman, M. L. (2001). Introduction. *Academy of Management Review, 26,* 507–511.

Gregersen, H. B., & Black, J. S. (1990). A multifaceted approach to expatriate retention in international assignments. *Group Organization Studies, 15*(4), 461–485. https://doi.org/10.1177/105960119001500409.

Gupta, B., Singh, D., Jandhyala, K., & Bhatt, S. (2013). Self-monitoring, cultural training and prior international work experience as predictors of cultural intelligence—A study of Indian expatriates. *Organizations and Markets in Emerging Economies, 4*(1), 56–71.

Hair, J. F., Black, W. C., Babin, B. J., Anderson, R. E., & Tatham, R. L. (2010). *Multivariate data analysis* (7th ed.). New Jersey: Pearson Prentice Hall.

Harzing, A. W. (2001). An analysis of the functions of international transfer of managers in MNCs. *Employee Relations, 23,* 581–598.

Hechanova, R., Beehr, T. A., & Christiansen, N. D. (2003). Antecedents and consequences of employees' adjustment to overseas assignments: A meta-analytic review. *Applied Psychology, 52,* 213–236.

Huang, T., Chi, S., & Lawler, J. J. (2005). The relationship between expatriates' personality traits and their adjustment to international assignments. *International Journal of Human Resource Management, 16*(9), 1656–1670.

Huff, K. C. (2013). Language, cultural intelligence and expatriate success. *Management Research Review, 36*(6), 596–612.

Huff, K. C., Song, P., & Gresch, E. B. (2014). Cultural intelligence, personality and cross-cultural adjustment: A study of expatriates in Japan. *International Journal of Intercultural Relations, 38*(1), 151–157.

Jyoti, J., & Kour, S. (2015). Assessing the cultural intelligence and task performance: Mediating role of cultural adjustment. *Cross-Cultural Management: An International Journal, 22*(2), 236–258.

Jyoti, J., & Kour, S. (2017a). Cultural intelligence and job performance: An empirical investigation of moderating and mediating variables. *International Journal of Cross Cultural Management, 17*(3), 305–326.

Jyoti, J., & Kour, S. (2017b). Factors affecting cultural intelligence and its impact on job performance: Role of cross-cultural adjustment, experience and perceived social support. *Personnel Review, 46*(4), 767–791.

Jyoti, J., Kour, S., & Bhau, S. (2015). Assessing the impact of cultural intelligence on job performance: Role of cross-cultural adaptability. *Journal of IMS Group, 12*(1), 23–33.

Kayworth, T., & Leidner, D. (2003). Organizational culture as a knowledge resource. In C. W. Holsapple (Ed.), *Handbook on knowledge management* (Vol. 1: knowledge matters, pp. 235–252). Berlin: Springer.

Kim, K., & Slocum, J. W. (2008). Individual differences and expatriate assignment effectiveness: The case of U.S.-based Korean expatriates. *Journal of World Business, 43*(1), 109–126.

Kumar, N., Rose, R. C., & Subramaniam (2008). The effects of personality and cultural intelligence on international assignment effectiveness: A review. *Journal of Social Science, 4*(4), 320–328.

Lee, L. Y. (2010). Multiple intelligence and the success of expatriation: The roles of contingency variables. *African Journal of Business Management, 4*(17), 3793–3804.

Lee, L. Y., & Kartika, N. (2014). The influence of individual, family and social capital factors on expatriate adjustment and performance: The moderating effect of psychology contract and organizational support. *Expert Systems with Applications, 41*, 5483–5494.

Lee, L. Y., & Sukoco, B. M. (2010). The effects of cultural intelligence on expatriate performance: The moderating effects of international experience. *The International Journal of Human Resource Management, 21*(7), 963–981.

Little, T. D., Card, N. A., Bovaird, J. A., Preacher, K. J., & Crandall, C. S. (2007). *Structural equation modeling of mediation and moderation with contextual factors* (pp. 207–230). Lawrence   Erlbaum   Associates.   Available   at   https://www.google.co.in/url?
sa=t&rct=j&q=&esrc=s&source=web&cd=1&ved=0CCIQFjAA&url=http%3A%2F%
2Fwww.quantpsy.org%2Fpubs%2Flittle_card_bovaird_preacher_crandall_2007.
pdf&ei=v8LMU9eyIMu7uASLu4HQBA&usg=AFQjCNHVlQi7-tf6VXa07ZZPcyEDubZxZ-
w. Accessed 21 May 2014.

MacNab, B. R., & Worthley, R. (2012). Individual characteristics as predictors of cultural intelligence development: The relevance of self-efficacy. *International Journal of Intercultural Relations, 36*, 62–71.

MacNab, B., Brislin, R., & Worthley, R. (2012). Experiential cultural intelligence development: Context and individual attributes. *International Journal of Human Resource Management, 23*, 1320–1341.

Malek, M. A., & Budhwar, P. (2013). Cultural intelligence as a predictor of expatriate adjustment and performance in Malaysia. *Journal of World Business, 48*(2), 222–231.

McDermott, R., & O'Dell, C. (2001). Overcoming cultural barriers to sharing knowledge. *The Journal of Knowledge Management, 5*(1), 76–85.

Mehra, N., & Tung, N. S. (2017). Role of adjustment as a mediator variable between cultural intelligence and well-being. *Journal if the Indian Academy of Applied Psychology, 43*(2), 286–295.

Moon, T. (2010). Emotional intelligence correlates of the four factor model of cultural intelligence. *Journal of Managerial Psychology, 25*(8), 876–898.

Moon, H. K., Choi, B. K., & Jung, J. S. (2012). Previous international experience, cross-cultural training and expatriates' cross-cultural adjustment: Effects of cultural intelligence and goal orientation. *Human Resource Quarterly, 23*(3), 285–330.

Naumann, E. (1993). Antecedents and consequences of satisfaction and commitment among expatriate managers. *Group Organization Management, 18*(2), 153–187.

Ng, K. Y., & Earley, P. C. (2006). Culture and intelligence: Old constructs, new frontiers. *Group Organization Management, 31*(1), 4–19.

Nonaka, I. (1994). A dynamic theory of organizational knowledge creation. *Organization Science, 5*(1), 14–37.

Paik, Y., & Shon, J. D. (2004). Expatriate managers and MNC's ability to control international subsidiaries: The case of Japanese MNCs. *Journal of World Business, 39*, 61–71.

Peltokorpi, V., & Froese, F. J. (2012). The impact of expatriate personality traits on cross-cultural adjustment: A study with expatriates in Japan. *International Business Review, 21*(4), 734–746.

Peng, M. W., Denis, Y. L. W., & Yi, J. (2008). An institution-based view of international business strategy: A focus on emerging economies. *Journal of International Business Studies, 39*, 920–936.

Podsakoff, P. M., MacKenzie, S. B., Lee, J. Y., & Podsakoff, N. P. (2003). Common method biases in behavioral research: A critical review of the literature and recommended remedies. *Journal of Applied Psychology, 88*, 879–903.

Preacher, K. J., & Hayes, A. F. (2004). SPSS and SAS procedures for estimating indirect effects in simple mediation models. *Behavior Research Methods, Instruments & Computers, 36*(4), 717–731.

Puck, J., Kittler, M., & Wright, C. (2008). Does it really work? Re-assessing the impact of predeparture cross-cultural training on expat adjustment. *International Journal of Human Resource Management, 19*(12), 2182–2197.

Ramalu, S. S., Wei, C. C., & Rose, R. Dr. (2011). The effect of cultural intelligence on cross-cultural adjustment and job performance amongst expatriates in Malaysia. *International Journal of Business and Social Science, 2*(9), 59–71.

Ramalu, S., Rose, R. C., Kumar, N., & Uli, J. (2010). Doing business in global arena: An examination of the relationship between cultural intelligence and cross-cultural adjustment. *Asian Academy of Management Journal, 15*(1), 79–97.

Ramayah, T., Yeap, J. A. L., & Ignatius, J. (2014). Assessing knowledge sharing among academics: A validation of the knowledge sharing behavior scale (KSBS). *Evaluation Review, 38*(2), 160–187.

Shaffer, M. A., & Harrison, D. A. (1998). Expatriates' psychological withdrawal from international assignments: Work, non-work, and family influences. *Personnel Psychology, 51*(1), 87–118.

Shaffer, M. A., Harrison, D. A., & Gilley, K. M. (1999). Dimensions, determinants, and difference in the expatriate adjustment process. *Journal of International Business Studies, 30*(3), 557–581.

Shay, J. P., & Baack, S. (2006). An empirical investigation of the relationships between and degree of expatriate adjustment and multiple measures of performance. *International Journal of Cross Cultural Management, 6*(3), 275–294.

Sohail, M. S., & Daud, S. (2009). Knowledge sharing in higher education institutions: Perspectives from Malaysia. *VINE: The Journal of Information and Knowledge Management Systems, 39* (2), 125–142.

Takeuchi, R., & Chen, J. (2013). The impact of international experience for expatriates' cross cultural adjustment: A theoretical review and critique. *Organization Psychology Review, 3*(3), 248–290.

Takeuchi, R., Yun, S., & Tesluk, P. E. (2002). An examination of crossover and spillover effects of spousal and expatriate cross-cultural adjustment on expatriate outcomes. *Journal of Applied Psychology, 87*(4), 655–666.

Takeuchi, R., Tesluk, P. E., Yun, S., & Lepak, D. P. (2005). An integrative view of international experience. *Academy of Management Journal, 48*(1), 85–100.

Thomas, D. C., Elron, E., Stahl, G., Ekelund, B. Z., Ravlin, E. C., Cerdin, J. L., et al. (2008). Cultural intelligence: Domain and assessment. *International Journal of Cross Cultural Management, 8*(2), 123–143.

Triandis, H.C. (2006). Cultural intelligence in organizations. *Group Organization Management, 3* (1), 20–26.

Tsang, E. W. K., & Kwan, K. M. (1999). Replication and theory development in organisational science: A critical realist perspective. *Academy of Management Review, 24*(4), 759–780.

Turner, C.H. & Trompenaars (2006). Cultural intelligence: Is such a capacity credible? *Group Organization Management, 3*(1), 56–63.

Van Dyne, L., Ang, S., & Livermore, D. (2010). Cultural intelligence: A pathway for leading in a rapidly globalized world. In K. Hannum, B. B. McFeeters, & L. Booysen (Eds.), *Leading across differences* (pp. 131–138). San Francisco: Pfeiffer.

Wang, M. (2016). Effects of expatriates' cultural intelligence on cross-cultural adjustment and job performance. *Revista de Cercetare si Interventie Sociala, 55,* 231–243.

Chapter 10
Employer Branding Analytics and Retention Strategies for Sustainable Growth of Organizations

Ravindra Sharma, S. P. Singh and Geeta Rana

Abstract Disruptive trends continue to create opportunities for organizations to quickly develop new capabilities and gain a competitive advantage. Employer branding and organizational attractiveness have garnered considerable research attention over the years owing to their significance in disruptive economy. Digitalization, global development, technological advancements, and greater dependence on data analytics have significantly accelerated market disruption, causing difficulties for employers in attracting employees in competitive advantages. For organizations, it is necessary to remain competitive; to this end, employers do a number of exercises to retain and attract employees. The aim of the research paper empirically explores the impact of organization's branding on employer attractiveness in Indian companies. Thus, 300 employees employed in various companies in India were surveyed. This paper uses correlation technique, factor analysis, and stepwise regression techniques to establish the impact of employer branding analytics on organizational attractiveness. Results suggest that branding analytics positively and significantly relates to companies' attractiveness. This paper offers deeper insights into the link between both of the variables, makes association between aspects and dimensions of the aforementioned constructs, and in doing so, provides significant implications for both researchers and practitioners. Findings of the study could help practitioners identify employer branding dimensions influencing organizational attractiveness the most. Practitioners could, with such knowledge, incorporate the most influential dimensions of employer branding in organizational culture.

Keywords Employer branding · Organizational attractiveness · Talent retention
Disruptive decade · Knowledge economy

R. Sharma (✉)
Uttarakhand Technical University, Dehradun, India
e-mail: ravindrasharma@srhu.edu.in

R. Sharma · G. Rana
Swami Rama Himalayan University, Dehradun, India

S. P. Singh
Gurukul Kangri Vishwavidyalaya, Haridwar, India

© Springer Nature Singapore Pte Ltd. 2019
H. Chahal et al. (eds.), *Understanding the Role of Business Analytics*,
https://doi.org/10.1007/978-981-13-1334-9_10

10.1 Introduction

Disruption is the latest buzzword in the business; it is an "event that results in a displacement or discontinuity." "Disruption" generally refers to the introduction of novel technologies' blockchain, virtual reality, and innovative models of business. In order to remain competitive when facing progressive acceleration in disruption, several firms are forced to reconsider and redesign their operations and offerings. The convergence of consumer expectations and novel technologies, in combination with cost and performance pressures, only increases the degree of difficulty. This makes present times both exciting and nervous for employers (Birkinshaw and Markides 2017). Firms today face several ongoing and potential threats to growth. Digitalization, global development, advances in technology, and higher dependence on data analytics have significantly accelerated market disruption, causing difficulty for employers in attracting employees in competitive advantages. For organizations, it is necessary to remain competitive; to this end, employers do a number of exercises to retain and attract employees (Figurska and Matuska 2013).

Disruption has become a growing concern among employers, not only in terms of how it affects their business model, but also the way it impacts the workforce and their skills. These evolving conditions are forcing employers to rethink their approach toward retaining and attracting employees. Employer branding (EB) is developing as a best practice to ensure that companies are prepared for an unpredictable future (Jiang and Iles 2011). In these disruptive times, nobody can claim complete control of their business and its readiness for what lies ahead. But one thing that remains true is that people are a company's most valuable asset. Any employer that can develop, retain, and attract great people will have the intellectual capital to compete far into the future, and this is the essence of contemporary employer branding (Bakanauskienė et al. 2014).

According to Ambler and Barrow (1996) EB in terms of an organizational culture (considering the organization as an employer) of communication and development. It suggests psychological, functional, and economic benefits offered by the employing firm, projecting it as a good workplace. Sullivan (2004) sees employment branding as a long-term strategy aimed at managing perception and awareness (of existing and potential employees and stakeholders) regarding a firm. Chhrabra and Mishra (2008) describe EB as a firm's identity creation and image management process as an employer. EB has been employed to project a firm as a superior employer, as well as incorporate techniques to engage and motivate employees. Van Mossevelde (2014) highlighted that lack of skilled labor and changed perceptions of the new generation were responsible for the emergence of EB. CIPD (2009) states that employer branding comprises intangible qualities and attributes that draw potential workers toward a firm. Backhaus and Tikoo (2004) emphasized that EB is a distinctive facet of a firm's employment offerings that differentiates it from rival firms. King and Grace (2008) stated that EB was the identity of an organization as an employer.

Another construct that has been taken up in the study is organizational attractiveness (OA); it is considered a competitive advantage for the employer and enables employees to feel comfortable and remain in the organization (Cable and Turban 2001). Robertson et al. (2005) observed that companies' attractiveness mediates between intention to agree to the job offer and recruitment message. According to Albinger and Freeman (2000), increasing media influence caused job seekers to become more cautious; any negative or positive information regarding a firm impacted potential employees' decision, influencing the recruitment process in turn. Carless and Imber (2007) asserted that OA referred to particular policies that attracted employees toward organizations. Bhatnagar and Srivastava (2008) opined that organizations should guard their attractiveness as employers so as to be able to attract talent amidst great competition in the business world. Ehrhart and Ziegert (2005) emphasized that OA greatly influenced whether workers stayed with the firm. Cable and Edwards (2004) said that OA increased by person-organization fit and it also helps potential candidates to compare their organizational needs, values, and personalities; the better the match more the probability of potential candidate to join the organization. According to Fombrun and Shanley (1990), it is vital for organizations to develop a positive corporate persona that appeals to potential workers as it boosts OA perception. Collins and Stevens (2002) stated that OA acted as important element in the job market which differentiates among potential employees (Armstrong 2007).

Disruptive changes across a variety of sectors significantly affect OA among employees. Unlike majority of previous studies that focus more on the link between different dimensions of EB and OA, the present paper seeks to examine the two constructs in such a way that EB's impact (as a whole and the effect of each aspect of EB) on each dimension of OA may be established. Such examination would offer deeper insights into the link between the two constructs and help academics and practitioners recognize specific aspects that affect particular dimensions more, and if any particular aspects have a greater role in determining OA. Thus, this study empirically analyzes the link between EB and OA using two approaches: (1) by regarding the two constructs as whole and (2) examining the association between OA constructs with different measurements of EB at different workplaces of India. Such a twofold analysis has not been attempted before in regard to the study variables and associations considered in this study, making the present piece of work a unique one. The association between separate aspects of OA and EB has been observed by previous studies; however, till date no study found the link between both constructs (as whole) while also analyzing the relationships between separate dimensions of OA and aspects of EB. Hence, this study carries immense originality value and significantly contributes to the extant literature.

10.2   Literature Review

Branding analytics helps organization employee for their development and retention policies while engaging employees in challenging task (Backhaus and Tikoo 2004). Authors examined what role EB played in retaining and attracting workers. Findings indicated that an organization established its own identity through its culture, style of management, quality of existing employees and economic conditions, social development, etc. Makwana and Dave (2014) examined EB practices in an Indian IT company. They emphasized that EB comprised beliefs and ideas that influenced how existing and potential workers perceived an organization, and the employment experience offered by the firm. Dawn and Biswas (2010) stated that quality and number of applicants could be increased by employing an effective employer branding strategy. Effective employer brand image development is a process that is gradual and warrants the creation of an organizational culture begetting a quality work environment meeting the expectations of potential workers. Hence, careful EB management is vital to its effectiveness. Lyons and Marler (2011) emphasized that in slowdown economy, companies must attract and provide opportunities in job market through the use of public network to attract capable employees.

According to Rai (2012), social media rose to global prominence rapidly, and technological characteristics influenced OA in a company. Ilesanmi (2014) examined the relevance and role of employer branding in retaining and attracting workers in disrupting decade. Even amid fierce competition, successful companies maintain their ability to draw and retain skilled workers by way of branding analytics which possesses higher capability and competencies for existing and new entrant's employees. Wilden et al. (2010) suggested that EB influenced employee decision to enter, leave, or remain with the organization. Shivaji (2013) opined that employer brand comprised a firm's "employment experience which includes factors like salary, rewards and benefits, organizational culture, management style as well as growth opportunities" and attracted employees toward the organization even in disruptive economy. Mak and Sockel (2001) proposed enforcement of policies for better career development, decreasing burnout and stress, fair remuneration, motivating workers, and creating a better organizational culture to decrease turnover tendency. According to CIPD (2014), robust EB associates HR policies, people strategy, and firm values to firm brand. Perception toward a brand as employer generates value that draws talent which, in turn, can be converted into profits (Copenhagen Business School 2009). Armstrong (2007) observed that individuals were drawn to firms that met their needs. According to Kapoor (2010), positive firm image developed by way of employer branding steadily drew applicants to the firm and conveyed that it was great to work in the firm.

Krishnan (2014) explained that employer value proposition referred to indirect and direct advantages that employees sought such as respect, diverse growth opportunities, empowering performance, and forward looking. Martin (2011) emphasized that EB identified with its missions, strategies, and cultures which attracted and retained talented employees in a competitive marketplace. Peyron (2013) argued that the younger generation was affected by aspects different from career growth and salary in a company. In a disruptive environment, jobs are viewed not just as means to money or career, but they are seen as a part of life. Thus, firms are expected to create an organizational environment, wherein workers may discover life and career paths. Highhouse et al. (2003) examined organizational attraction as surrogate assessment of organizational pursuit.

This scarcity of the literature highlights the need for the present study. The present study analyzes the relevance of branding analytics in companies' attractiveness in Indian companies, especially in disrupting economy. Thus, this study seeks to determine if a firm's employer brand acts as a significant factor when deciding to join or remain in the firm.

The discussion above suggests that EB affects OA. Branding values mentioned above such as economic, interest, application, salary, security, development capability bear similarity to many EBs included in the present study. This paper aims to determine the association between EBs and OA in the context of India and, in doing so, fill the gap in the extant literature. To achieve this objective, links between separate aspects of EB and dimensions of OA have been analyzed while observing the overall affect of EB on overall OA. It is therefore proposed:

H1 Significant association exists between employer branding ("social value, development value, application value, interest value, and economic value") and organizational attractiveness ("general attractiveness, intention to pursue, and prestige"). Also, employer branding will significantly predict organizational attractiveness (Fig. 10.1).

10.3 Methodology

10.3.1 Sample

Three hundred employees of different Indian companies were approached "for the purpose of data collection using convenience sampling." Questionnaires containing questions on OA and EB were used to collect data. Recorded information about individual such as name, gender, occupation, highest qualification, designation, work experience, and family status was sought through questionnaires. Table 10.1 is used to describe and summarize the data.

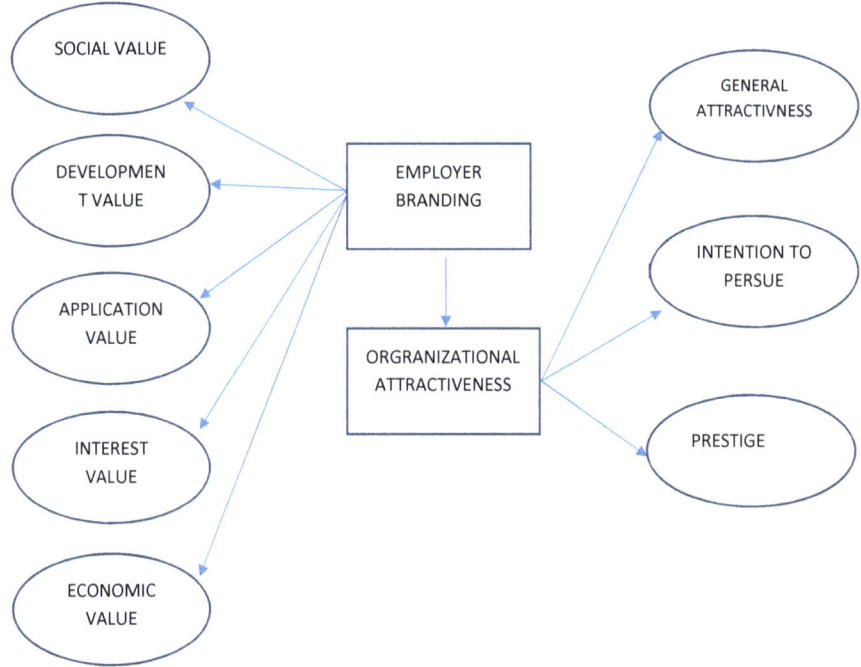

Fig. 10.1 Proposed model

10.3.2 *Instruments*

Data were collected with the help of two measuring instruments. A description of each scale has been given below.

EB instrument carries 25 items generated by Berthon et al. (2005). The scale identifies five dimensions. These are "social value, development value, interest value, economic value, and application value." The use of the instrument in various nations and contexts, and across occupational levels establishes its validity. The reliability coefficients for these dimensions are 0.91, 0.91, 0.89, 0.91, and 0.91, respectively. Respondents were requested to reply on a seven-point Likert scale ("anchored on 'to a very little extent' and 'to a very great extent'").

The dimensions of employer branding have been defined below:

1. **Interest value** assesses the degree of attraction of a person toward an employer providing a work environment that is exciting, new work practices, and that uses creativity of employees to manufacture innovative and high-quality services and products.
2. **Social value** observes the degree of attraction of a person toward an employer providing a happy and fun working environment that presents healthy collegial associations and a team atmosphere.

Table 10.1 Distribution of frequency values in the sample (*n* = 300)

Demographic variable	Frequency	Percent
Gender		
Male	274	91.3
Female	26	8.7
Total	300	100
Marital status		
Married	268	89.3
Unmarried	32	10.7
Total	300	100
Educational qualification		
Graduate	166	55.3
Postgraduate	134	44.7
Total	300	100
Age		
25–40	114	38
41–56	144	48
57–72	42	14
Total	300	100
Work experience		
1–15	146	48.7
16–30	131	43.7
31–45	23	7.6
Total	300	100

3. **Economic value** assesses how much a person is drawn toward organization salary compensation package, promotional opportunities, and security.
4. **Development value** evaluates the degree of attraction of a person toward "an employer that offers confidence, self-worth, and recognition along with career-enhancing experience and a launch pad to employment in the future."
5. **Application value** "assesses the degree to which a person is drawn toward an employer" offering workers opportunities to apply their learning and educate others in a humanitarian- and customer-oriented environment.

10.4   Organizational Attractiveness Scale (OAS)

OA was assessed on a 15-item scale given by Highhouse et al. (2003). The scale identifies three dimensions. "These are *general attractiveness, intentions to pursue, and prestige.*" The reliability coefficients for these dimensions are 0.77, 0.92, and 0.95, respectively. Respondents were requested to reply to a "five-point Likert scale (1 = *strongly disagree*; 5 = *strongly agree*)."

Dimensions of employer branding have been defined below:

1. **General attractiveness**. Initial attitudes toward the firm as potential employer.
2. **Intentions to pursue**. Assessed intentions toward the company with a "modern approach to transacting with the firm in the future."
3. **Prestige**. Emphasis on facets of a firm exposed to "social influence such as status, popularity, and reputation."

10.5 Statistical Analysis

Various statistical analyses such as correlation and stepwise regression techniques were carried out on collected data to assess the influence of employer branding on organizational attractiveness. Along with these data analysis techniques, factor analysis is used.

10.6 Total Variance Explained

Table 10.2 shows the factor matrix for both the study variables. Employer branding scale was subjected to factor analysis on the basis of principal factor analysis; the three factors were obtained up to eigenvalues over 1.00, and they explained 53.69% of total variance. Factor analysis was carried out on employer branding scale; 23 of 25 items were kept for further analysis; items having more than 0.55 factor loadings were chosen for the study. Communalities giving proportion of variance for each original variable are put in the last factor matrix column. Table 10.2 shows rotated factor solutions.

Table 10.2 Employer branding scale rotated component matrix

Variables/items	Components			h^2
	1	2	3	
SV1	0.63			0.43
SV2			0.79	0.65
SV3	0.74			0.62
SV4		0.68		0.54
SV5			0.83	0.69
DV1	0.62			0.46
DV2		0.70		0.53

(continued)

Table 10.2 (continued)

Variables/items	Components			h^2
	1	2	3	
DV3		0.78		0.62
DV4	0.66			0.58
DV5		0.74		0.59
AV1		0.66		0.55
AV2	0.49[a]			0.29
AV3	0.78			0.63
AV4	0.67			0.47
AV5	0.64			0.55
IV1	0.61			0.38
IV2		0.42[a]		0.46
IV3	0.69			0.51
IV4			0.63	0.42
IV5		0.59		0.38
EV1			0.65	0.45
EV2	0.72			0.35
EV3		0.63		0.45
EV4			0.62	0.47
EV5			0.58	0.27
Eigenvalues	6.76	4.78	4.10	12.34
Percentage of variance	25.882	17.249	10.238	53.369

Source Authors' own
Notes Score less than 0.55 was deleted from the solution
[a]SV, social value; DV, development value; AV, application value; IV, interest value; EV, economic value

For organizational attractiveness, 12 items out of 15 were obtained and 2 items were eliminated from the scale owing to factor loadings below 0.55. The scale accounted for 65.41% of variance, and the communalities ranged from 0.48 to 0.98. Table 10.3 presents rotated factor solutions.

Table 10.3 Organizational attractiveness scale rotated component matrix

Variables/items	Components			h^2
	1	2	3	
GA1	0.80			0.67
GA2			0.98	0.98
GA3		0.74		0.68
GA4		0.83		0.70

(continued)

Table 10.3 (continued)

Variables/items	Components			h^2
	1	2	3	
GA5		0.69		0.65
IP1	0.63			0.56
IP2	0.77			0.68
IP3	0.74			0.64
IP4	0.59			0.49
IP5		0.53[a]		0.48
P1		0.54[a]		0.55
P2	0.63			0.63
P3		0.74		0.71
P4		0.62		0.69
P5			0.65	0.45
Eigenvalues	7.14	1.05	1.63	9.56
Percentage of variance	51.05	7.54	6.81	65.41

Source Authors' own
Notes Item scores less than 0.55 were deleted from the solution
[a]GA, general attractiveness; IP, intentions to pursue; P, prestige

The criteria mentioned above permit the entry of five predictors—"development value, social value, interest value, economic value, and application value." Collectively, these dimensions contribute to the judgment of dimensions of organizational attractiveness.

10.7 Analysis

Clearly, robust correlations between every independent and dependent variable support our hypothesis. Table 10.4 shows correlation between both the variables displayed through a significant link between the two with the value $r = 0.78**$ ($p < 0.01$ level). The association between EB and OA (on overall basis) is shown with the help of a graph (see Fig. 10.2). Among all the dimensions of EB, that is "social value, development value, application value, interest value, and economic value were significant for the general attractiveness, intention to pursue, and prestige" (Table 10.5).

Stepwise multiple regression (see Table 10.5) shows that of all employer branding dimensions, economic value predicted general attractiveness with multiple R as 0.63 ($F = 69.42**$, $p < 0.01$, $\beta = 0.63$, $R^2 = 0.40$), and social value with multiple R as 0.65 ($F = 38.61**$, $p < 0.01$, $\beta = 0.21$, $R^2 = 0.43$), and jointly accounted for 40% variance in the prediction of general attractiveness. All together, economic value emerged as the most robust predictor of general attractiveness with estimated beta value of 0.63.

Table 10.4 Intercorrelation, mean, and standard deviation between constructs of EB and OA ($N = 300$)

S. No.	Variables	Mean	S.D.	1	2	3	4	5	6	7	8
1.	Social value	2.10	0.5712	1							
2.	Development value	2.27	0.5433	0.88**	1						
3.	Application value	2.17	0.5282	0.90**	0.86**	1					
4.	Interest value	2.13	0.5433	0.54**	0.51**	0.52**	1				
5.	Economic value	2.38	0.5832	0.42**	0.56**	0.43**	0.59**	1			
6.	General attractiveness	2.35	0.4692	0.50**	0.47**	0.45**	0.85**	0.62**	1		
7.	Intention to pursue	2.25	0.6058	0.56**	0.67**	0.51**	0.71**	0.61**	0.67**	1	
8.	Prestige	2.35	0.6107	0.52**	0.57**	0.71**	0.61**	0.66**	0.77**	0.69**	1

Source Authors' own

** It dipicts values are singnificant at .05 level

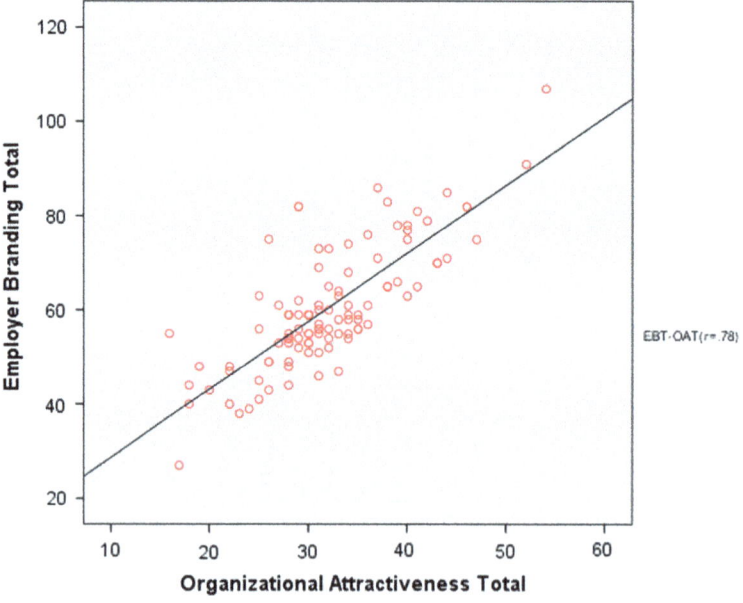

Fig. 10.2 Relationship between EBT and OAT *Source* Authors' own

Table 10.5 Stepwise regression analysis

Variables	R	R^2	SEM	F-value	D.F.	β-value
D.V.: general attractiveness						
Economic value	0.63	0.40	0.3644	69.42	1298	0.63
Economic value, social value	0.65	0.43	0.3574	38.61	1297	0.49, 0.21
D.V.: intention to pursue						
Application value	0.70	0.49.	0.4336	100.92	1298	0.70
Application value, development value	0.75	0.56	0.4034	17.17	1297	0.41, 0.39
Application value, development value, social value	0.76	0.58	0.3973	4.19	1296	0.34, 0.31, 0.19
D.V.: prestige						
Economic value	0.64	0.41	0.4706	72.84	1298	0.64
Economic value, social value	0.69	0.48	0.4408	15.53	1297,	0.40, 0.36
Economic value, social value, application value	0.71	0.51	0.4336	4.43	1296	0.30, 0.27, 0.22

Source Authors' own

The organizational attractiveness dimension intention to pursue was predicted by the dimension of employer branding application value R as 0.70 ($F = 100.92^{**}$, $p < 0.01$, $\beta = 0.70$, $R^2 = 0.49$); development value with estimated R as 0.75 ($F = 17.17^{**}$, $p < 0.01$, $\beta = 0.39$, $R^2 = 0.56$) and social value with the calculated

R as 0.76 ($F = 4.19^{**}$, $p < 0.01$, $\beta = 0.19$, $R^2 = 0.58$), and jointly accounted for 58% variance in the prediction of intention to pursue. All together, application value emerged as the most robust predictor of intention to pursue with estimated beta value of 0.70.

Finally the organizational attractiveness dimension prestige was predicated on the basis of employer branding dimensions and economic value predicted prestige with multiple R as 0.64 ($F = 72.84^{**}$, $p < 0.01$, $\beta = 0.64$, $R^2 = 0.41$); social value with multiple R as 0.69 ($F = 15.53^{**}$, $p < 0.01$, $\beta = 0.36$, $R^2 = 0.48$) and application value with the multiple R as 0.71 ($F = 4.43^{**}$, $p < 0.01$, $\beta = 0.22$, $R^2 = 0.51$), and jointly explained 51% variance in the prediction of prestige. All together, economic value emerged as the most robust predictor of prestige with estimated beta value of 0.41. Thus, based on results obtained, Hypothesis 1 is retained at 0.01 level.

10.8 Discussion

10.8.1 Relationship Between EB and OA

Table 10.4 shows that EB leads to general attractiveness, intention to pursue, and prestige. Findings indicate that employer branding provides a reputation to the firm as the best place to work; employers have to offer economic, functional, psychological, and social benefits to "attract and retain the best talent," and by doing so, firms can enhance employee motivation (Dell et al. 2001; Copenhagen Business School 2009). EB is the perfect package that provides salary, proper atmosphere, and career development opportunities to their employee which attract them toward organization. According to Verma and Verma (2014), employers are responsible to provide workers the perfect workplace. Sullivan (2004) suggested that employer branding was an approach for organizations during a slowdown in the economy; it helped build an image in potential workers' minds of "a grade place to work" (Minchington 2010). Social value, development value, and economic value at the workplace provide envisioned benefits to workers so that they attract toward organizations. Development value and application value motivate employees to effactually utilize their skills and abilities at work, and lead them to perceive the workplace as enjoyable; this gives shape to their expectations with respect to their employment (Lievens 2007).

Table 10.4 shows that EB dimensions positively correlate with general attractiveness, intention to pursue, and prestige. When employees perceive the workplace as interesting, socially supportive, and filled with creative value, they undergo growth and advancement which causes employee confidence and satisfaction; further, they are simultaneously attracted toward the organization. Findings further suggest that development, application, and economic values encourage employees' satisfaction, high ROI on professional as well as personal levels (Dawn and Biswas 2010).

Results indicate that employer branding in disrupting economic time conveys a firm to generate a choice of employer.

10.8.2 Prediction of OA with the EB Dimensions

Table 10.5 clearly indicates that economic value is the most robust predictor of all three EB dimensions. General attractiveness was predicted by economic and social values. Findings imply that economic value at place of work, such as above average salary, job security, and career development opportunities, attracts employees in a competitive market. In disruptive market, employees are generally less likely to pursue alternative employment; they prefer companies offering secure jobs and promotional opportunities for continuous learning (Wallace et al. 2014).

Social value attracts employees toward the employer and recognizes the people in troubled economic times. In economic downturn, employers provide progressive, friendly, and enjoyable environment to attract employees (Berthon et al. 2005; Ambler and Barrow 1996). Application value, development value, and social value also positively attract employees which delivers a stimulating innovative work environment along with building a set of different competencies among employees through training, coaching, and mentoring opportunities (Minton-Eversole 2009; Cooper 2008). All these values collectively provide better career development and greater challenges a modern approach to transacting with the firm in the future (intention to pursue). Economic value, social value, and application value also emerged as key predictors of prestige. It has been suggested that these dimensions of employer branding change the way employees think by developing creative programs and provide challenges, recognition, and empowerment to retain and attract brightest employees in disruptive economy. Economic value, social value, and application value also emerged as key predictor for prestige. It has been suggested that these dimensions of employer branding change the way employees think by developing creative programs and provide challenges, recognition, and empowerment to retain and attract brightest employees in disruptive economy. Findings indicate that economic value influences a company's image in terms of reputation, popularity, and status (prestige); therefore, employees attract toward healthy climate of innovation which engages and retains employees.

10.9   Implications and Contribution

This study carries significant implications for practitioners and academics. Findings reveal that EB influences OA; thus, managements could use our findings to recognize EB aspects that are particularly effectual in obtaining OA. It has been observed that "economic value, application value, social value and development value emerged as strong predictors of attracting and retaining employees in recent trend such as downsizing, and outsourcing", "employers provide employees with marketable skills through training and development in return for effort and flexibility (Baruch 2004)". Practitioners, with some effort, would be able to recognize the dimensions helpful in drawing workers to firms, thus leading to OA.

10.10 Limitations and Future Research Direction

In this research paper, limitation is as follows: First, the population could have been better represented by a larger sample. Second, employees at only senior managerial and managerial positions have been considered. Validating the findings of this study with the help of larger samples could be an area of future research. Further, similar studies could be conducted in multicultural organizational contexts and environments.

10.11 Conclusion

This paper empirically analyzes the link shared by EB and OA. This is done by considering both constructs as whole, as well as examining the association between the different dimensions of EB and aspects of OA. Such twofold analysis makes the present piece of work unique, contributing significantly to the extant literature. Results indicate that OA (as a whole) is influenced by EB (as a whole). Further, "social value, economic value, application value, and development value" are OA dimensions commonly influencing three EB aspects, suggesting that these aspects are vital to OA. Economic value appeared as the most robust predictor as it affected all three OA aspects. Employers could create work environments encouraging and enforcing job security with many other packages of benefits such those career development branding values that provide culture and experiential benefits to the employees.

References

Albinger, H. S., & Freeman, S. J. (2000). Corporate social performance and attractiveness as an employer to different job seeking populations.

Ambler, T., & Barrow, S. (1996). The employer brand. *Journal of Brand Management, 4,* 185–206.

Armstrong, M. (2007). *Armstrong's handbook of human resource management practice*. London Kogan Page.

Backhaus, K., & Tikoo, S. (2004). Conceptualizing and researching employer branding. *Career Development International, 9,* 501–517.

Bakanauskienė, I., Lina, Ž., & Justina, V. (2014). Employer's attractiveness: Employees' expectations vs. reality in Lithuania. *Human Resources Management & Ergonomics, 8,* 23–37.

Baruch, Y. (2004). *Managing careers: Theory and practice.* Prentice Hall.

Berthon, P., Ewing, M., & Erthen, K. (2005). Captivating company: Dimensions of attractiveness in employer branding. *International Journal of Advertising, 24,* 151–172.

Bhatnagar, J., & Srivastava, P. (2008). Strategy for staffing: Employer branding & person organization fit. *The Indian Journal of Industrial Relations, 44,* 35–48.

Birkinshaw, J., & Markides, C. (2017). *Dealing with disruption how is the digital disruption affecting your business? Today's podcast theme is dealing with disruption.* London Business School. https://www.london.edu/faculty-and-research/lbsr/iie-digital-disruption-podcast-one#. Wg0ccluCzIU.

Cable, D., & Edward, M. (2004). The determinants of job seekers' reputation perceptions. *Journal of Organizational Behaviour, 21,* 929–947.

Cable, D. M., & Turban, D. B. (2001). Establishing the dimensions, sources, and value of job seekers' employer knowledge during recruitment. In G. R. Ferris (Ed.), *Research in Personnel and Human Resources Management, 20,* 115–163.

Carless, S. A., & Imber, A. (2007). Job and organizational characteristics. A construct evaluation of applicant perceptions. *Educational and Psychological Measurement, 67,* 328–423.

Chhrabra, H., & Mishra, D. (2008). New Zealand talent acquisition and employer branding case studies. *Human Resources Magazine, 17,* 26–27.

CIPD. (2009). *Employer brand.* London: CIPD.

CIPD (2014). Employer brank—factsheets—CIPD. Available at https://www.cipd.co.uk/hr-resources/factsheets/employer-brank.aspx (online).

Collins, C. J., & Stevens, C. K. (2002). The relationship between early recruitment-related activities and the application decisions of new labour market entrants: A brank equity approach to recruitment. *Journal of Applied Psychology, 87,* 1121–1133.

Cooper, K. (2008). *Attract, develop and retain: Initiatives to sustain a competitive workforce.* Spring Hill, Qld: Mining Industry Skills Centre.

Copenhagen Business School (2009). Employer branding—A possible value creating attribute Available at http://employerbrandingmarketing.wordpress.com/thesis.

Dawn, K. S., & Biswas, D. (2010). Employer branding: A new strategic dimension of Indian corporations. *Asian Journal of Management Research, 7,* 21–33.

Dell, D., Ainspan, N., & Bodenberg, T. (2001). *Engaging employees trough your brand* (Research Report No. 1288-01-RR). New York: The Conference Board.

Ehrhart, K. H., & Ziegert, J. C. (2005). Why are individuals attracted to organizations? *Journal of Management, 31,* 901–919.

Figurska, I., & Matuska, E. (2013). Employer branding as a human resources management strategy. *Human Resources Management & Ergonomics, 8,* 36–48.

Fombrun, L. M., & Shanley, I. (1990). What's in a name? Reputation building and corporate strategy. *Academy of Management Journal, 33,* 233–258.

Highhouse, S., Lievens, F., & Sinar, E. (2003). Measuring attraction to organizations. *Educational and Psychological Measurement, 63,* 986–1001.

Ilesanmi, O. (2014). *The relevance of employer branding in attracting and retaining employees in Nigeria's brewery industry.* Available at http://esource.dbs.ie/bitstream/handle/10788/1808/mba_ilesanmi_o_2014.pdf?sequence=1. Accessed on 12/03/2017.

Jiang, T., & Iles, P. (2011). Employer-brand equity, organizational attractiveness and talent management in the Zhejiang private sector; China. *Journal of Technology Management in China, 6,* 97–110.

Kapoor, V. (2010). Employer branding: A study of its relevance in India. *IUP Journal of Brand Management, 7,* 51–75.

King, C., & Grace, D. (2008). Building and measuring employee-based brand equity. *European Journal of Marketing, 44,* 938–971.

Krishnan, S. (2014). The importance of employer branding, Deccan Herald, 28th Mar. Available at http://www.deccanherald.com/content/109808/importance-employer-branding.html.

Lievens, F. (2007). The relation of instrumental and symbolic attributes to a company's attractiveness as an employer. *In Personnel Psychology, 56,* 75–102. https://doi.org/10.1111/j. 1744-6570.2003.tb00144.

Lyons, B. D., & Marler, J. H. (2011). Got image? Examining organizational image in web recruitment. *Journal of Management Psychology 26,* 58–76.

Mak, B. L., & Sockel, H. (2001). A confirmatory factor analysis of IS employee motivation and retention. *Information and management 38,* 265–276.

Makwana, K., & Dave, G. (2014). Employer branding: A case of Infosys. *International Journal of Humanities and Social Science Invention, 3,* 42–49.

Martin, G. (2011). Is there a bigger and better future for employer branding? Facing up to innovation, corporate reputations and wicked problems in SHRM. *International Journal of Human Resource Management, 22,* 3618–3637.

Minchington, B. (2010). *The employer brand manager's handbook: Torrensville: Collective learning.*

Minton-Eversole, T. (2009). Quality measurement: Key to best-in-class talent acquisition. *HR Magazine, 6,* 64–66.

Peyron, C. (2013). The quest for purpose, employer branding today. Available at http://www. employerbrandingtoday.com/. Last accessed 28 Mar 2014.

Rai, S. (2012). Engaging young employees (Gen Y) in a social media dominated world—Review and retrospection. *Procedia-Social and Behavioral Sciences, 37,* 257–266.

Robertson, Q., Collins, C., & Oreg, S. (2005). The effects of recruitment message specificity on applicant attraction to organizations. *Journal of Business and Psychology, 19,* 31–39.

Shivaji, W. S. (2013). Employer branding: A strategic tool to attract and retain talents in a competitive environment. *Indian Streams Research Journal, 2*(12), 1–4.

Sullivan, J. (2004). Eight elements of a successful employment brand. ER daily, 23 Feb. Available at www.erexchange.com/artilces/db/52CB45FDADFAA4CD2BBC366659,E26892A.asp.

Van Mossevelde, C. (2014, March 14). What-is-employer-branding. Retrieved from http:// universumglobal.com/2014/03.

Verma, D., & Verma, C. (2014). A study on attractiveness dimensions of employer branding in technical educational institutions. *Journal of Human Resource Management, 6,* 36–43.

Wallace, M., Lings, I., Roslyn C., & Sheldon, N. (2014). Attracting and retaining staff: The role of branding and industry image. In *Workforce development.* Singapore: Springer Science + Business Media. https://doi.org/10.1007/978981-4560-58-0_2,19.

Wilden, & Gudergan, S., Berth, K. (2010). Employer branding: Strategic implications for staff recruitment. *Journal of Marketing Management, 26,* 56–73.

Lightning Source UK Ltd.
Milton Keynes UK
UKHW02n1142180918
329106UK00002B/18/P